VENOMOUS EARTH

VENOMOUS EARTH

How arsenic caused the world's worst mass poisoning

Andrew A. Meharg

Macmillan

First published 2005 by
Macmillan
Houndmills, Basingstoke, Hampshire RG21 6XS and
175 Fifth Avenue, New York, N. Y. 10010
Companies and representatives throughout the world

ISBN-13: 978–1–4039–4499–3
ISBN-10: 1–4039–4499–7

This book is printed on paper suitable for recycling and made from fully managed and sustained forest sources.

A catalogue record for this book is available from the British Library.

A catalog record for this book is available from the Library of Congress.

10 9 8 7 6 5 4 3 2 1
14 13 12 11 10 09 08 07 06 05

Printed and bound in China

He gathered all that springs to birth
From the many-venomed earth;
First a little, thence to more,
He sampled all her killing store;

A. E. Housman, *Terence, this is stupid stuff*, from *A Shropshire Lad*, 1896

Für Caroline

CONTENTS

LIST OF FIGURES

ACKNOWLEDGEMENTS

It was the mines of Devon that fuelled my interest in arsenic, and I am grateful to Mark Macnair at the University of Exeter for giving me gainful employ to work on them. Joerg Feldmann at Aberdeen is thanked for many chats on the history of arsenic. There are many scientists who have done pioneering work on the arsenic problems of Bangladesh and elsewhere, and I have tried to highlight their work accurately in the text. Any factual errors in interpreting their research are mine.

My research into William Morris's arsenic connections has been greatly aided by Peter Cormack at the William Morris Gallery, Walthamstow, who provided me with the first piece of Morris & Co arsenic-saturated wallpaper and allowed me to sift through his archive. Michael Parry of Sanderson Wallpaper Company allowed me generous access to the Morris & Co. collection that his company owns. Michael has also granted permission to reproduce Morris & Co. patterns such as *Indian* (used on the cover) and *Trellis* (reproduced in the text). These patterns and others are still available from Sanderson's, although the use of arsenic in Morris & Co. papers ceased in the 1870s! The staff at the Victoria and Albert Museum also responded very generously to my peculiar requests to look at and analyse various parts of their building as well as their Morris & Co. and Jeffrey & Co. archives.

David Kinniburgh of the British Geological Survey provided very helpful comments on the manuscript.

Many thanks go to Sara Abdulla for commissioning and editing this book – and to Tom Clarke for his part in initiating the project.

Shahid and Nilufa Islam Hossain were my kind and generous hosts in Bangladesh, showing me their beautiful country. Thanks also to Imamul Huq for his help during my stay in Dhaka.

I thank my parents Maureen and Billy for their love and support, and Kenny, Christine and Annie for providing fun throughout. And to Caroline, my love, my joy.

A.A.M.

Permissions

The author and publisher are grateful for permission to reproduce extracts from the following works:

Boyle, R. *The Works of Robert Boyle* edited by M.H. Hunter and E.B. Davis. Published by Pickering & Chatto, London, 1999. Reproduced with permission.

The British Medical Journal. The BMJ Publishing Group grants permission for quotes to be reproduced with non-exclusive world rights in print and electronic formats for this and all future editions of this Work.

Chasteen, T.G., Wiggli, M. and Bentley, R. Of garlic, mice and Gmelin: the odor of trimethylarsine. *Applied Organometal Chemistry*, **16**(6): 281–286 June 2002. © John Wiley & Sons Limited. Reproduced with permission.

Graves, R. Translation of the *Twelve Caesars* by Suetonius. Due acknowledgement is made of the permission of A P Watt Ltd on behalf of The Trustees of the Robert Graves Copyright Trust.

Housman, A.E. *Terrence, this is stupid stuff*. Reproduced with permission of The Society of Authors as the Literay Representative of the Estate of A. E. Housman.

Morris, W. *The Collected Letters of William Morris*, Vol. II. Reproduced with permission from Princetown University Press.

Levi, P. *The Periodic Table*. Reproduced with permission by Penguin Books UK, and Knopf Publishing Group, USA.

Pliny the Elder, *Natural History*. Reproduced with permission by Penguin Books UK.

Every effort has been made to trace the copyright holders but if any have been inadvertently overlooked the publishers will be pleased to make the necessary arrangement at the first opportunity.

Chapter 1
THE DEVIL'S WATER

Ours truly is a God-forsaken country. Difficult, indeed, is it for us to maintain the strength of will to do. We get no help in any real sense.

Rabindranath Tagore, *Glimpses of Bengal*, Cuttack, February 1893

The puncturing of the Earth's skin with tens of millions of wells has drawn out poison – not just any poison, but the most notorious of all: arsenic. Over one hundred million people may now be at risk of routinely drinking dangerous levels of arsenic, an element that causes skin, bladder and lung cancer, stillbirths and heart attacks.

'A month ago', wrote the *New York Times* in November 1998, 'young mother Pinjira Begum found out that her own slow dying was nothing unusual, that tens of thousands of Bangladeshi villagers are suffering the same ghastly decay, their skin spotted like spoiled fruit and warts and sores covering their hands and feet'. The mother of three had been abandoned by her husband for a second wife. 'She was pretty once, but now she is too thin and smells bad and is uglier by the day', he explained. Said Pinjira: 'They tell me the water has made me sick, but others use this well and they are not sick'. Her ten-year-old daughter already had the tell-tale spotting of arsenic poisoning on her chest. Her husband was fine.

Pinjira died the following May. Her dying wish was 'Please save my three children from the deadly arsenic'.

o–o

Water that was meant to bring life has brought death. It is a turn of events cruel beyond belief. The peoples of the crowded delta plains of south-east Asia affected by this crisis are among the poorest on Earth. They suffer floods, cyclones, famine and disease. A populace could scarcely be less capable of coping with poison in its drinking water: arsenic has no taste and no immediate side effects, and it took decades to realise that the most insidious of mass poisonings was occurring. Not a year passes without millions more being added to the list of potential victims.

The situation is worst in the Bengal Delta, which encompasses the People's Republic of Bangladesh and Indian state of West Bengal. The most conservative estimates reckon that five million people in West Bengal and 35 million in Bangladesh are being poisoned. At the other extreme, estimates suggest that many more may be affected: the United Nations World Health Organization (WHO) has called the disaster 'the largest mass poisoning of a population in history' and estimates that up to 77 million people are potentially affected in Bangladesh alone. A report published in 2000 in the *Bulletin of the WHO* by the epidemiologist Professor Allan Smith, of the University of California, states: 'The scale of the environmental disaster is greater than any seen before; it is beyond the accidents in Bhopal, in 1984, and Chernobyl, Ukraine, in 1986'.

o-o

The ancient Indian state of Bengal encompasses the world's largest delta, comprising the convergent Ganges, Brahmaputra and Meghna rivers. Its peoples are predominantly Muslim; Hindus form the largest minority, with much smaller Buddhist and Christian populations. The Bengali language is spoken uniformly across the Delta, crossing religious and cultural divides. This sets Bengal apart from the Urdu-speaking Muslim populations to the west of the Indian subcontinent, in Pakistan.

On independence from Britain, India divided along religious grounds, despite Mahatma Ghandi's opposition and considerable distress. Hindu-dominated West Bengal was incorporated into modern India, and Muslim-dominated East Bengal was renamed East Pakistan. East Pakistan was separated by a subcontinent from West Pakistan, 1,000 miles away.

Unsurprisingly, Pakistan was riven with strife from its inception. East Pakistan, with 56% of the country's population, was poorer than West Pakistan. Bengali Muslims were not overly favoured in the Civil Service under Indian rule from London. Muslims tended to have a less well developed educational infrastructure than Hindus, with higher rates of illiteracy. As the Hindu civil servant class fled Pakistan on independence from

Britain, the small Urdu-speaking population of East Bengal and the Urdu-speaking middle classes from West Pakistan dominated the civil and military structures in East Bengal. East Pakistan generated over half of Pakistan's wealth, mainly through jute exports, but it received just one third of this wealth back, as the factory owners tended to be West Pakistani. In 1971 civil war broke out. East Bengal, assisted by the Indian army, achieved independence later that year and was renamed Bangladesh.

The West not only stood back and watched the persecution and genocide of Hindus and Bengali separatists by West Pakistan's military; they also armed Pakistan. The West was playing a bigger game. It considered Pakistan to be a bulwark against 'Red' China, and was willing to ignore the humanitarian catastrophe in East Bengal caused by Pakistan suppressing Bengali nationalism. The situation was exacerbated by cyclones that killed between 300,000 and a 500,000 people in the Bay of Bengal in 1970, leaving the survivors without food, clean water or housing. Only limited and late assistance came from the West, cautious not to upset the government of Pakistan by interfering in East Pakistan.

India did not want war either. It feared the humanitarian upheaval and the millions of refugees; eventually 10 million came. They also feared Bengali nationalism wanting a reunification of West and East Bengal. India asked the USA to stop supplying Pakistan with arms. America agreed in public, but a leaked memo from Henry Kissinger to President Nixon stated 'tilt in favour of Pakistan'. The leak was the last straw, and India entered the war for independence on the side of the Bengalis. Bangladesh was left to hang due to the brutal politics of the Cold War. As in all civil wars, all sides suffered greatly.

Bangladesh survived and has built itself into a proud nation; with a few hiccups, it has rapidly evolved into a secular democracy. It has battled with fortitude despite its many problems, most of which were not of its own making. Routine one-day strikes, known as 'hartals', are used by whichever party is in opposition to try to destabilise the government, but there is a fundamental sense of fairness.

Centuries of oppression and religious conflict meant that this new nation had few natural or human resources to deal with its enormous and impoverished population. The war of 1971 left the infrastructure in tatters. One of the most densely populated countries in the world, Bangladesh could not grow enough food to support its own population. A collapsed economy and monsoon floods led to a terrible famine – a famine that shocked the world, with its images of starving children with distended stomachs, their eyes active only with swarming flies. As these images beamed into comfortable Western homes, Bangladesh became synonymous with starvation and desperation.

Aid eventually started to pour in, first for famine relief and then to improve the infrastructure to try to prevent future disasters. This improvement had to start with eradicating the diseases that kept the country in misery: the water-borne killers diarrhoea, typhus and cholera.

Today the Bangladesh economy is starting to take off due to a cheap and skilled labour force. The images of the 1970s' famine and war, and too much indifferent travel writing, fail to capture the vibrancy of this culturally rich and enormously bountiful place. Poverty, though, is all around: the wealth of the land breeds overpopulation.

o-o

Bangladesh is one of the world's wettest countries. It is criss-crossed by a multitude of rivers with ever-changing paths. As the meeting point of the great Ganges and the Brahmaputra, it has been called the plughole of the Indian subcontinent. Around 70% of its surface is routinely flooded during the monsoon season between June and September. This yearly engorgement of its waterways massively reworks the landscape, destroying roads and bridges.

Why is such a difficult landscape so densely populated? The rivers carry massive sediment loads, billions of tonnes, that make the soils of Bangladesh extremely fertile. Intensive agriculture following the Green revolution has fed enormous population growth. New, heavier-cropping rice varieties, particularly those that thrive through the dry season, have led to much greater food production. In a land where your security against old age is children, more food means more births. In many ways, Bangladesh has been a victim of its own agricultural success.

But the life-giving waters mobilise human effluent, spreading disease. The monsoons effectively make the country a connected sewer. Luke Harding, describing a visit to Bangladesh's capital, Dhaka, for the *Guardian* in 2002, wrote: 'This Dhaka – probably the most squalid, wretched, and perversely beautiful city on earth'. One quarter of Dhaka's 12 million people live in slums, in unimaginable filth. He describes one of these slum townships centred on an ox-bow lake. 'All slum-dwellers are forced to use the lake as a latrine. They use the same water to clean their cooking pots, to wash clothes, and to bathe.' Dhaka has only one sewage treatment plant, which is often broken.

In the 1970s nearly one quarter of a million died each year in Bangladesh through water-borne disease. Pure drinking water supply was a necessity. Salvation seemed at hand in the sediments swept into the delta plain of Bengal. Over the millennia these have built up a vast repository of pure water pouring down from the Himalayas, free of surface contamination. This water could be readily harvested using cheap tubewell technology, where iron tubes are pushed through the soft sediment to tap water stores below. The tops of these tubes are fitted with a hand pump to draw water to the surface.

The earliest records of tubewells in Bangladesh stretch back to 1937, with slow rates of construction until 1970. In 1972 a massive tubewell drilling operation was instigated by the United Nations Children's Fund (UNICEF), which installed 900,000 such devices throughout Bangladesh and encouraged others to sink them. The initial health benefits of the well water were rapidly apparent. Infant mortality halved from 151 per thousand in 1960 to 83 per thousand in 1996. Over the same period the mortality rate for under-fives dropped from 247 per 1000 to 112 per 1000. Today, most of the approximately 10 million tubewells in Bangladesh are privately owned and sunk, up to 94% in one study, with the remainder supplied either by the community, such as at mosques and bazaars, or by aid agencies, such as UNICEF. By 1997 UNICEF announced that it had surpassed its target of supplying 80% of Bangladeshis with safe drinking water by the year 2000. Around 95% of rural inhabitants now have access to hand-pump tubewells.

West Bengal shares the same delta plain as Bangladesh, and suffers from similar overcrowding, poverty and disease. UNICEF-sponsored tubewell sinking also occurred widely, here but statistics on tubewells are harder to find.

Not all of the people of the Bengal Plain were happy with the tubewell programme. 'Devil's water is coming! Devil's water is coming!' the locals cried in response to the first extraction of groundwater in the Nadia District of West Bengal during the 1960s. Local lore held that water from the ground was tainted by the devil. Aid agencies spent considerable effort convincing people drinking surface waters to switch to tubewells.

And so another disaster, not caused by wind or rain, not carried by rivers or the barrel of a gun, was released.

o-o

The annals of the crisis state that arsenic was first detected in groundwaters in 1978. The reference is obscure, as it is only briefly cited, without context, in subsequent studies. The first concrete report of arsenic contamination of tubewell water was in 1983. Dr Saha, a dermatologist at the School of Tropical Medicine in Calcutta (Kolcuta), the capital of West Bengal, saw patients from nearby villages with mysterious spotted skin on their trunks, arms and legs. The spots looked like black raindrops. The patients also had rough skin and warts on their palms and soles. His investigation led him to the village of Gangapur in West Bengal. Saha took tubewell samples for testing. He found that the waters contained arsenic.

The first indication of illness caused by long-term arsenic exposure, termed arsenicosis, is a black raindrop pattern on the limbs, chest and back (Figure 1.1), and sometimes on mucus membranes such as the tongue and gums. This is called hyperpigmentation. White spots on the skin, hypopigmentation,

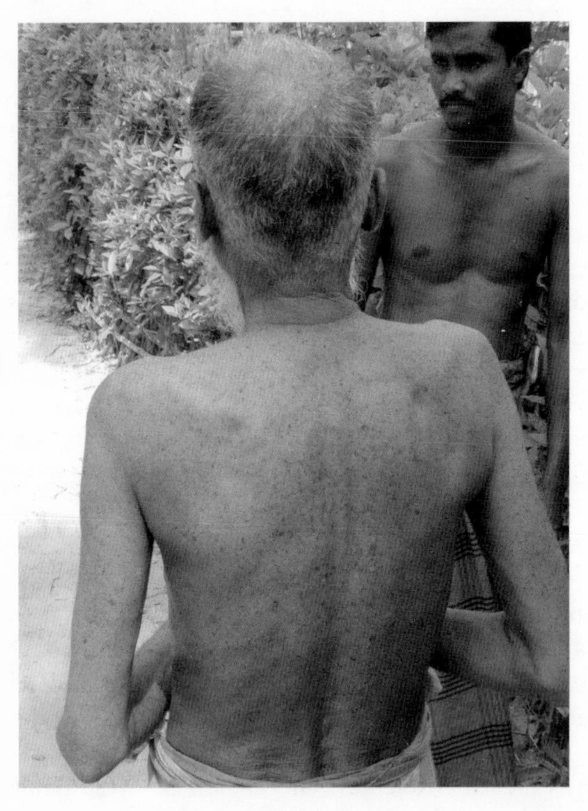

Figure 1.1 Black rain.

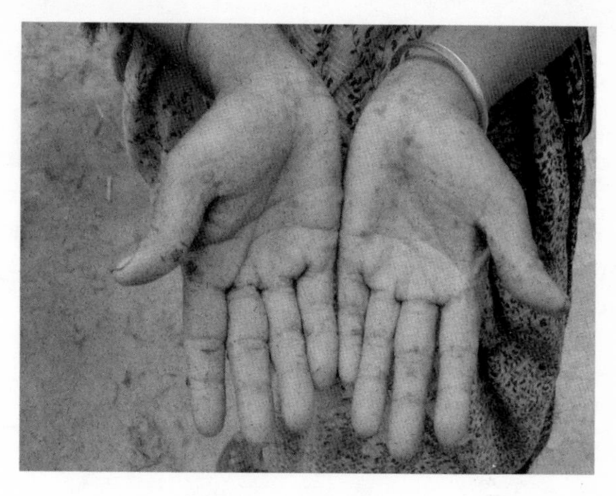

Figure 1.2 Keratosis.

can also occur. Both of these symptoms are also known as melanosis. The youngest patients with hyperpigmentation are two years old. Typically, after 10 years' exposure the skin on the hands and feet hardens and rises into nodules. These can reach up to 1 cm across. This stage of arsenic poisoning is called keratosis (Figure 1.2). With even more exposure, skin cancer develops on the keratosis sites. The speed at which arsenic-related skin diseases arise and progress depends on the dose.

Arsenic can cause other skin diseases. It is associated with 'black foot' or peripheral vascular disease. This ultimately results in gangrene in the extremities. This is not widely seen in Bangladesh, but is typical of long-term arsenic exposure in areas of Inner Mongolia and Taiwan. The reasons for these regional differences in how the body responds to the poison are not known.

Over long periods, exposure to arsenic damages the skin, kidneys, brain, heart, and circulation; miscarriages and stillbirths also seem to increase. But bladder and lung cancers are the major killers. How arsenic makes us sick, and in particular how it causes cancer, is not clear. All we know is that it interferes with the workings of our enzymes and genetic material.

o-o

Arsenic compounds are deadly at high concentrations. For the chemical compound arsenic trioxide, the deliberate poisoner's choice, 70 to 180 milligrams kill. A teaspoon of arsenic trioxide in a glass of wine was the Borgias' trick. But as far as groundwater goes, more pertinent is: what is the lowest dose that results in chronic injury?

Table 1.1 Common forms of arsenic.

Name	Chemical formula	Form	Other names
Arsenic	As	Silvery metallic solid which quickly tarnishes to dark grey or black	
Arsenate	AsO_4^{3-}	White solid as sodium and potassium salts	
Arsenite	AsO_3^{3-}	White solid as sodium and potassium salts	
Monomethyl arsonic acid	$H_2As(CH_3)O_3$	White solid	MMA
Dimethylarsinic acid	$HAs(CH_3)_2O_2$	White solid	DMA, cacodylic acid
Arsine	AsH_3	Colourless gas	
Trimethyl arsine	$As(CH_3)_3$	Colourless gas	Gosio's Gas
Arsenobetaine	$HAs(CH_3)_3CH_2CO_2$	White solid	Fish arsenic
Realgar	AsS	Red solid	*Sandarach*
Orpiment	As_2S_3	Yellow solid	*Arsenikon*
Arsenic trioxide	As_2O_3	White solid	White arsenic, *Hütten-rauch*, *Hidri*
Copper arsenite	$CuAsHO_3$	Green solid	Scheele's green
Copper acetoarsenite	$3CuO.As_2O_3.Cu[OOC.CH_3]$	Green solid	Emerald green

In terms of environmental pollutants there is nothing unusual in arsenic's acute toxicity. Metal salts of mercury, copper and cadmium, for example, are similarly deadly. Many modern pesticides (HCH, aldrin, DDT, PCP) are also as acutely poisonous as arsenic salts. What really sets arsenic apart is that it is a carcinogen that is often naturally elevated in the environment. This natural occurrence is then exacerbated by a host of industrial activities.

It is, therefore, very important to assess long-term chronic exposure of arsenic. Most experts agree that 10 micrograms of arsenic, in the soluble inorganic forms known as arsenite or arsenate, in a litre of water is 'safe'. However, like radiation, the toxicity of inorganic arsenic is linear, so damage to the body is caused at even lower concentrations. So this 'safe' level is a concentration that has an 'acceptable risk' for human health, a pragmatic balance between health effects and environmental exposure. Arsenic is present throughout the environment, usually at low concentrations. A truly safe dose would be zero.

Ten micrograms of arsenic in a litre of water is equivalent to 10 grams of arsenic in one billion grams of water – environmentalists use the term 10 ppb meaning '10 parts per billion'. This is roughly analogous to one third of a teaspoon of arsenic dissolved in an Olympic-sized swimming pool.

The value 50 ppb is considered 'safe' by the government of Bangladesh. The risk of getting cancer from 20 years of drinking arsenic at 50 ppb is one in a hundred. How did this unsafe level of arsenic come to be considered OK?

In the early twentieth century 50 ppb was derived to be the safe concentration for acute – not chronic – exposure to arsenic. In the city of Manchester, England, in 1900, contaminated sugar was used in brewing. The resulting beer contained 15,000 ppb arsenic, poisoning 6,000 people and causing 70 deaths. An enquiry found that a safe acute dose of arsenic in beer was 1,000 ppb. The enquiry arbitrarily decided that dividing this number by 20 should give a sufficient margin of safety.

This 50 ppb figure was adopted throughout the world. The UK, the USA and WHO all accepted it. It was only when epidemiological data from the arsenic aquifers of Taiwan started to emerge in the 1960s that people realised that this dose was problematic.

o-o

By 1987 Saha had identified 1,214 cases of chronic arsenical dermatosis directly related to the drinking water. These patients came from 61 separate villages in seven administrative districts. The problem appeared to be widespread in West Bengal. Saha and his colleagues published their findings in a series of papers in the *Indian Journal of Dermatology* from 1984 onwards.

The world was slow to react. Saha's studies only received a wider audience with a landmark paper in the journal *Bulletins of the World Health Organization* in 1988. *Bulletins*, as it is generally referred to, is the academic journal for considering the needs of developing nations; it often covers the supply of drinking water, improving sanitation, and fighting infectious disease. Moreover, it is published by WHO, an agency of the United Nations. It was UNICEF, don't forget, that facilitated the tubewell revolution in the Bay of Bengal and elsewhere.

The paper in *Bulletins* was by A.K. Chakraborty and co-workers. It was entitled 'Chronic arsenic toxicity from drinking tubewell water in rural West Bengal'. It told the following story. Eight men and five women from the village of Ramnagar, West Bengal, were admitted to the Institute of Post-Graduate Medical Education and Research in Calcutta with signs of chronic arsenical dermatosis. Ramnagar is 40 km south of Calcutta, on the Ganges delta. It is an agricultural area with no industrial infrastructure. The villagers were poor, living in mud huts or brick houses. Most were labourers; a few worked in offices. All obtained water from the same tubewell, which had 2,000 ppb of arsenic, 40 times higher than the 50 ppb drinking water standard in the region and 200 times higher than the now generally agreed safe level of 10 ppb.

Chakraborty's team went to the village and examined the relatives of one of their patients – 48 people in all. Of these, 46 had skin hyperpigmentation and liver damage. The two unaffected family members drank from another tubewell that had 200 ppb arsenic – 10 times lower than the water that the rest of their family were drinking.

What's more, the amounts of arsenic in nail and hair samples from the 13 people admitted to hospital were, on average, 100 times greater than in samples taken from people drinking water with less than 10 ppb arsenic.

This study should have set alarm bells ringing. It did not. The publication was marked by silence. The delay cost the Bengalis dear: tubewell digging proceeded at an exponential rate during the 1990s, exposing tens of millions more to arsenic-contaminated groundwater. Tubewells became such an important status symbol that they were often included in a bride's dowry. Better-off households often have two or more.

o–o

In 1993 concern was raised again by one Dipankar Chakraborti (no relation to A.K. Chakraborty), a Calcutta-based analytical chemist. Visiting his home village in West Bengal, Chakraborti saw that many villagers had the black rain of arsenicosis on their chests and limbs. This encounter spurred him to champion the study of arsenic pollution in West Bengal, and later Bangladesh. He and his colleagues at the School of Environmental Studies at Jadavpur University started to record the extent of arsenic contamination in the wells of West Bengal villages where arsenicosis was common. Chakraborti campaigned continuously: arranging conferences, travelling the world speaking to aid donors, and trying to explain to local governments. His words were generally dismissed as hype. 'I have been told that I am creating panic. But every new survey identifies more villages', he recalled in the mid-1990s. His pleas often fell on deaf ears. Aid officials and governments simply could not believe what they were hearing. Chakraborti's independently funded (from government or aid agencies) research into six districts in West Bengal predicted that nearly 200,000 of the region's 30 million inhabitants had arsenic-induced skin disease. Given that the real boom in tubewell digging occurred throughout the 1990s, and that the latency of arsenic-related skin disease is thought to be around ten years, the prognosis was grim.

o-o

Initially, geologists and other scientists considering the arsenic problems of West Bengal thought that the poison was coming from minerals in the Ganges valley, further upstream of West Bengal. These pyrites and arsenopyrites contain sulphur, iron and arsenic. A theory was developed in the late 1980s that the Ganges carried pyrites and arsenopyrites to the sediments of West Bengal. There, the arsenic in these pyrites was thought to dissolve when tubewell digging let oxygen-rich air into the aquifers. It was mostly Indian scientists that thought this, and it was a good educated guess. Unfortunately for the people of Bangladesh, who were told that the arsenic calamity was located in West Bengal, it was the wrong guess. The arsenic source was much more diffuse.

The first formal report of problems in Bangladesh came in 1993 after the discovery of an arsenic-affected village in 1992. The Department of Public Health and Engineering (DPHE) in Dhaka published a paper in the *Journal of Preventive and Social Medicine* saying that arsenicosis was now present in Bangladesh. The first high-arsenic tubewell was discovered near Nawabganj town, close to West Bengal. The poisoned wells were again traced back from the black rain of hyperpigmentation on the villagers drinking from them. An opportunity to mark the advance of the arsenic into Bangladesh was missed. No one followed up the observation. The government of Bangladesh was going through a period of instability at this stage, and the sufferings of ordinary people were sidelined.

Some claim that there was another missed opportunity in 1992. The United Kingdom's earth sciences research outfit, the British Geological Survey, proposed a project to the UK Department for International Development (DFID) to conduct a chemical survey of tubewell waters in Bangladesh because of the lack of any chemical data on Bangladesh aquifers at that time. DFID agreed to fund this project. The goals of the BGS were to investigate the suitability of groundwater for fish farming and to investigate deep tubewells, not to test for arsenic. They, like most others, did not know that there were high levels of arsenic in Bangladeshi aquifers. Because the team did not test for the presence of arsenic, the legal firm of Leigh, Day and Co., acting on behalf of 400 Bangladesh citizens, sued the BGS in the British High Court, an action funded almost entirely by British Legal Aid. A majority ruling by the Court of Appeal at the end of February 2004 eventually found in favour of the BGS. The judges decided that the BGS did not have a 'duty of care' to the villagers in the area surveyed. It is ironic that the BGS has been taken to court for its actions in Bangladesh, as its monumental arsenic tubewell survey of 1998 and 1999 is an exemplary study, providing hard data for management of the crisis. The BGS was asked to study deep tubewells which we now know are, in the main, free from arsenic contamination and are used for irrigation, not for drinking water. In fact, along with Dipankar Chakarborti, the BGS has been the main champion of the cause of the people of the Bengal Delta.

o-o

It was up to the West Bengali campaigner Dipankar Chakraborti to alert the world to a much greater disaster. It was soon apparent to him and others that the scale of the problem across the border in Bangladesh was much greater than that in West Bengal.

Bangladeshi patients started to drift into Calcutta clinics with signs of advanced arsenicosis. Once more, Chakraborti set off to investigate. Through the efforts of his research group, further arsenic contamination of tubewells in Bangladesh was confirmed by 1995.

The arsenic calamity started to become more widely known by scientists outside the Indian subcontinent following the Arsenic Conference in Calcutta in 1995, which was hosted by Chakraborti with the aim of informing the governments of Bangladesh and India and international aid agencies. In 1996, following the Calcutta conference, WHO declared the Bangladesh situation a 'Major Public Health Issue'.

Chakraborti, along with staff from the Dhaka Community Hospital, then began to survey Bangladesh to quantify the extent of the problem. The results were not published until 1997, 14 years after arsenicosis had first been observed in West Bengal. Chakraborti surveyed 294 tubewells in the Rajarampur Village of Nawabganj District, Bangladesh. He found that 29% of the tubewells had over 50 ppb arsenic. Another survey conducted by Chakaraborti during 1996 and 1997, in Samta village, Jessore District, Bangladesh, one of the worst affected areas, showed that over 90% of wells had arsenic levels higher than 50 ppb. By 1998 the government of Bangladesh had conducted around 100,000 tests for arsenic, but this data was not systematically compiled and presented. The most effective survey to date, covering most of Bangladesh, was conducted by the BGS and Bangladesh's DHPE in 1998 and 1999 and analysed 3534 wells. Over 25% had 50 ppb arsenic or more.

The BGS survey left out only the hill tracts of Chittagong, which are arsenic free, and the relatively uninhabited, and inaccessible mangrove forests of the Sundabans in the south. The northern territories, particularly the north-west corner of Bangladesh, are arsenic-free. Contamination is generally concentrated in the south. Of Bangladesh's 64 districts, 59 contain zones of arsenic-tainted groundwater. To date, no such survey exists of West Bengal.

A inspection by Dhaka Community Hospital and Chakraborti's School of Environmental Studies, Calcutta of 18 Bangladesh districts, found that out of 1,630 individuals examined, 57.5% had arsenic-related skin lesions. Children as young as 10 were afflicted. A larger study of 7,364 people found that one third had arsenic-related skin lesions. To date, 17,000 individuals have been recorded with arsenic-related disease in Bangladesh, although the figure may be higher due to the inaccessibility of rural populations, stigma surrounding arsenic disease, and failures in diagnosis. In the future, assuming 35 million Bangladeshis are drinking water with more than 50 ppb arsenic, a figure calculated by the BGS from its survey, and given that the United States Environmental Protection Agency has calculated that there is a one in a hundred risk of developing bladder or lung cancer from drinking 50 ppb arsenic, the prediction is that 350,000 people will develop fatal cancers. This is the minimum figure – the more arsenic people are exposed to, the greater the cancer risk, and levels up to 4,000 ppb have been recorded in the Bengal Basin. The Bangladesh and West Bengal medics are braced for an epidemic of cancers in the next 10 years.

Major social problems ensue from arsenic-related disease. For example, marriages are annulled and people with arsenicosis are avoided. In some areas, panic sets in. With so many likely to fall ill, a huge burden has been placed on family units and communities, and their land. The socio-economic fallout of the arsenic crisis could mirror that of the African AIDS epidemic, albeit on a smaller scale. But there is one difference between this groundwater crisis and an infectious disease: arsenic can be turned off at the tap if another tap with unpolluted water can be turned on.

o-o

Only as the enormity of the extent of the arsenic distribution in Bangladesh and West Bengal became apparent did other questions start to be asked. Dipankar Chakraborti first noticed in the mid-1990s that the amounts of arsenic in toenail and hair samples and in urine were much higher than could be accounted for by water intake alone. There had to be another dietary source of arsenic. This remained a puzzle for a number of years until I surveyed the paddy soils and rice grain of Bangladesh. I found that tubewells were poisoning food as well as drinking water. The poison was being liberally poured onto the most precious of all the delta's natural resources: its soil. Not only has the miracle of clean water been turned sour – the agricultural revolution is slowly poisoning the land.

Paddy rice (*Oryza sativa* L.), is the staple food of Bangladesh, providing 73% of its calorific intake. Around three-quarters of the total cropped area is given over to rice cultivation and more than four fifths of the total irrigated area is used for it. Vast quantities of tubewell water are pumped up to flood paddy fields during the long, hot dry season. Once arsenic is in the soil it generally stays there, unless the monsoon floods wash away the contaminated sediment.

Rice plants and vegetables, such as arum and bitter gourd, can readily accumulate soluble forms of arsenic into their edible parts. Rice grown in Bangladesh has been found with as much as 1,830 ppb arsenic; a normal level would be 200 ppb. Rice, it seems, can contribute almost as much arsenic to the diet as a tubewell – especially where it is also cooked in contaminated water, driving levels up even higher.

But if people stopped drinking tubewell water, they would still need it to irrigate paddy rice. The quantities of water required for cultivation are so vast that it would be uneconomical to rid the water of arsenic. Instead, clean water could be obtained from rainwater harvesting (Bangladesh is one of the wettest countries in the world) or through political negotiations with India to release more water into shared river systems. The latter option is not straightforward: the construction of dams to divert and regulate water flowing down the Ganges has been a source of considerable political tension since Bangladesh's war of independence. During the dry season India siphons off the precious Ganges waters; during the flood seasons it opens the gates, unleashing torrents of water into Bangladesh.

Even if the use of tubewell water stopped and if arable land were irrigated with river water, the arsenic already in the soil would remain. The cost of removing this, even if it were possible, would be exorbitant.

o–o

The global response to the arsenic crisis in the Bengal Delta has been marked by staggering inertia. The lack of action by the Bangladesh and Indian governments is part of the problem. The health issues in their sea of humanity, though, are overwhelming. AIDS, tuberculosis, malaria, diarrhoea, malnutrition – arsenic is just another entry on a long list of killers. Bangladesh, in particular, has also undergone substantial periods of political instability. But the lack of action on the side of the international aid community is more puzzling. They after all initiated and funded the tubewell digging, admittedly with the very best of intentions. Only because of the unstinting efforts of local scientists, mainly based in Calcutta, did the world finally sit up and listen.

The World Bank's *World Development Report 2004* makes distressing reading. 'While the number of individuals showing the symptoms of arsenic poisoning is still low – despite the high concentrations of arsenic in the water – between 25 and 30 million people may be at risk in the future.' At least 350,000 internal cancers are expected, never mind arsenic-related skin diseases. How many need to have this threat hang over them before the World Bank deems that a moderate or large number of people are exposed? The BGS estimates that 35 million people are drinking water containing 50 ppb or more of arsenic. WHO in 2000 estimated that 35–77 million are potentially exposed to too high a level of arsenic in their drinking water. (The upper figure is derived by adding up the population in the affected areas, not all of whom will be drinking water with more than 50 ppb.) *This* is the 'at risk' population. How did the World Bank reach its 'at risk' figure of 25–30 million in 2004? Why are the 77 million people said to be 'at risk' of drinking 50 ppb arsenic 'in the future'? They are either drinking it, or are at risk of drinking it, now. If they are drinking water with more than 50 ppb arsenic, they have at least a one in a hundred chance of getting cancer. And what about those now consuming water contaminated with arsenic above the safer level of 10 ppb? The BGS calculates that 50 to 60 million people are in this category. Worse, one well can have water high in arsenic while another 10 metres away is low in the element. This wide variation makes it difficult to assess accurately the total number of people in jeopardy.

The World Bank document admits: 'The first response to the unfurling problem by government and many donors was denial'. Kazi Matin Ahmed, a geologist working at the University of Dhaka, comments:

> These wells were introduced into the ground water in good faith and they've saved countless lives. But saying that, you could have rightly expected UNICEF to have known a lot sooner about the arsenic, especially after Dr. Dipankar Chakraborti had issued his warnings. At the very least, someone must be blamed for bad management of such a vast system of water supply.

UNICEF states: 'At the time, standard procedures for testing the safety of groundwater did not include tests for arsenic, which had never before been found in the kind of geological formations that exist in Bangladesh'. UNICEF in Dhaka says it is 'unproductive to dwell on the past'. In 1998 a UNICEF representative Shahida Azfar said: 'We are wedded to safe water, not tubewells, but at this time tubewells remain a good, affordable idea and our program will go on'.

Water is now tested for arsenic before any pump is installed. But these tests are not foolproof: in Bangladesh a well is considered safe if it has 49.9 ppb arsenic present. An extra 0.1 of a ppb, though, makes it 'unsafe'. And well water arsenic concentrations often vary by a few ppb or more over short periods of time.

Since 1998 up to 4 million tubewells have been sunk, each providing water to at least one family, if not several. Statistically, one million of these tubewells will be arsenic-contaminated, poisoning one million families. Timothy Claydon of WaterAid explains the stark choice for the citizens of the Bengal plain: 'It can sometimes be a choice between death by arsenic poisoning or death by diarrhoea'. Removing this gamble must be the focus of the aid agencies. Water free from both bacteria and arsenic needs to be provided.

Initial attempts to get an aid program organised stalled badly. In 1998 the World Bank gave the government of Bangladesh an interest-free loan of $34 million to fight the arsenic crisis. Two years later it remained largely unspent, with just $2 million used – on committees and consultants. By 2004 still only $6 million had been spent. It is now predicted that the $34 million will be spent by the end of 2005. Another $55 million is already lined up for the next loan to counteract arsenic in drinking water. Nothing at all is being done about the slow accumulation of arsenic in soil. Many more millions will need to be spent before the pollution is reversed, or at a minimum, stopped.

o-o

National and international aid efforts at mitigation have been, in the main, ineffective, even though the problem was identified over twenty years ago. In 2001, WHO argued that helping the people of Bengal to fight arsenicosis has been limited because: 'The poor availability of reliable information hinders action at all levels and may lead to panic, exacerbated if misleading reports are made. Effective information channels have yet to be established to those affected and concerned'.

Do we have enough information to act now? My answer is unequivocally yes. Each study shows the same patterns of contamination and the same patterns of disease. The Bengal arsenic poisoning is undoubtedly the greatest man-made chemical disaster in history. It will be decades before the full consequences are apparent given the time arsenic-induced diseases take to develop. Aid agencies and governments can no longer sit back and wonder what to do next. They must wean people off arsenic-contaminated tubewell water and find cheap, plentiful and truly safe alternatives.

o-o

In the late 1990s scientists started to realise that the geological conditions that have given rise to the problems in the Bengal Delta are not unique. Arsenic-contaminated groundwater could be present, they realised, in delta after delta around the coast of south-east Asia. We now also know that other conditions lead to high levels of arsenic in groundwaters in aquifers far inland.

As Figure 1.3 shows, large tracts of land in the arid parts of the USA, South America and Inner Mongolia, and volcanic regions such as New Zealand, all feature arsenic-contaminated waters. The legacy of base and precious metal mining throughout the globe has resulted in shocking levels of arsenic contamination of drinking water supplies, particularly in Thailand, Chile and China. Still, it is the estuarine regions of south-east Asia that are causing the most concern. Enormous populations of the world's poorest and most malnourished people crowd into these fertile plains. For example, arsenic has now been found further and further up the course of the Ganges. In the last couple of years the Indian state of Bihar in the Middle Ganga Plain was found to have aquifers with high levels of arsenic. Early estimates suggest that 200,000 people live in the area. The situation is worse further up the Ganges, with an estimated 10 million residents of the Terai Plain, Nepal, now thought to be at risk, with many showing

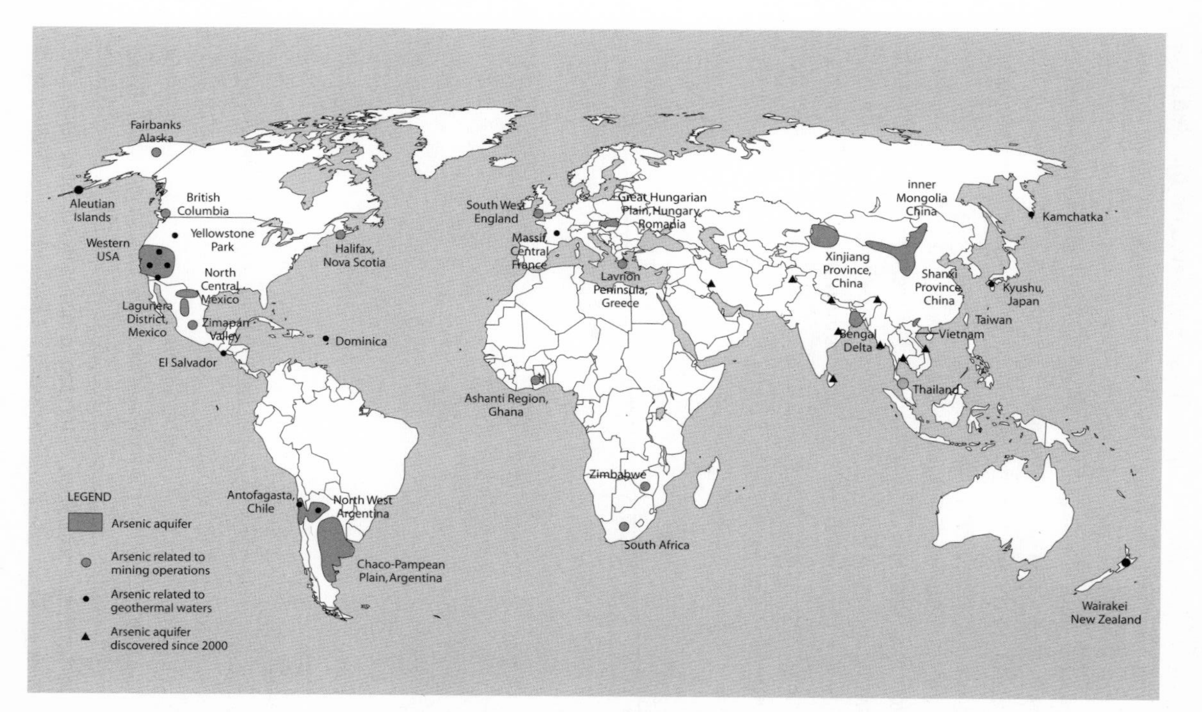

Figure 1.3 Map of the world's arsenic-rich aquifers and mining contaminated regions.

signs of arsenicosis. The city of Hanoi, situated on the Red River delta in Vietnam, has high levels of arsenic, with estimates suggesting that more than 10 million people are at risk. Tubewells were only sunk here in the mid-1990s – arsenicosis has not yet been detected in the population. The deltas of Myanmar, Cambodia and Laos have also been reported to be contaminated. Over 5 million Chinese are exposed to arsenic-laced groundwater. Recently, Iran and Pakistan have been added to this list of affected countries. The number of affected peoples is simply growing and growing now that we know where to look. Based on the best estimates so far, over 100 million people, or 1.5% of the planet's population, are living in areas with dangerous levels of arsenic in the drinking water. 'The Devil's water' abounds.

Chapter 2

A NATURAL DISASTER

If there were any beings in the nether world, assuredly the tunnelling brought about by greed and luxury would have dug them up. Are we surprised if the Earth has brought forth creatures to harm us? Wild animals, I believe, guard her and ward off impious hands. Do we not mine amongst the snakes...?

Pliny the Elder, translated by John F. Healy. *Natural History*, Penguin Classics, 1991

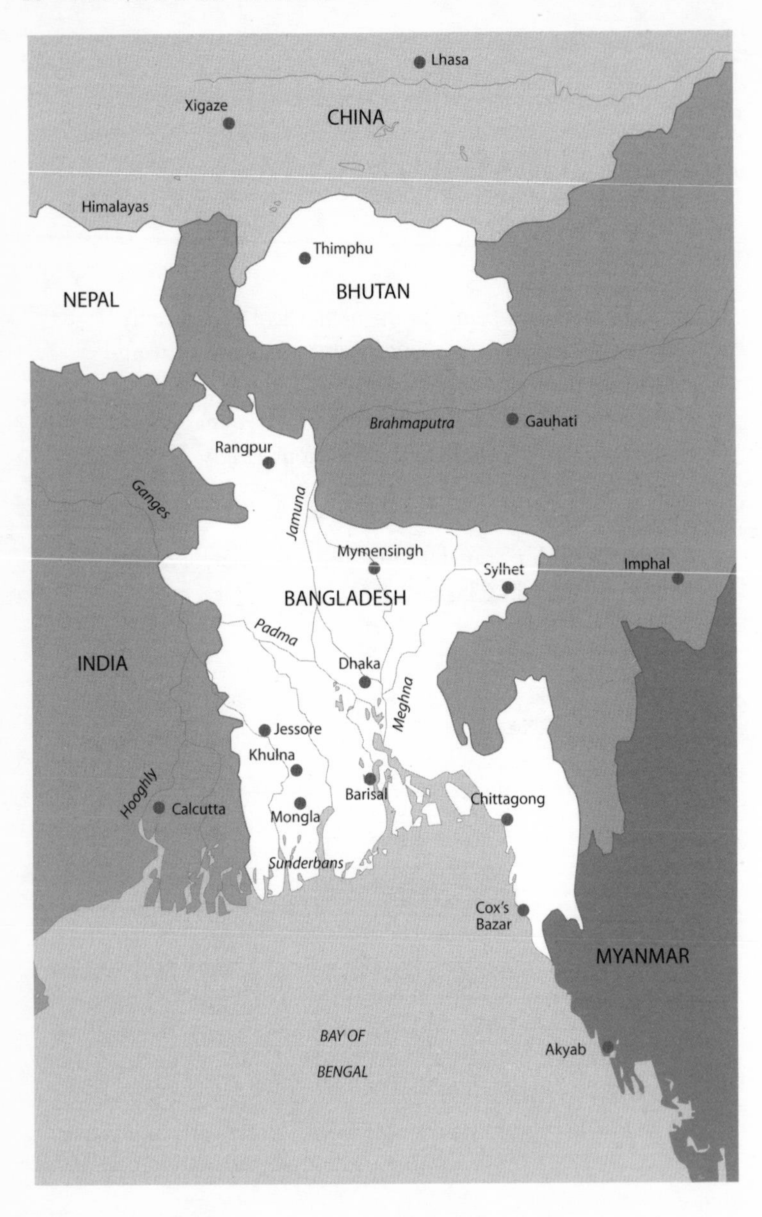

Figure 2.1 The rivers of the Ganges Plain.

Rivers dominate the Bengal Delta (Figure 2.1). It is a land of lush paddies and oxbow lakes, strung out like jewels among the silver ribbons of the Padma, Jamuna and Meghna rivers. Padma and Jamuna are Bangladesh's names for India's holy Ganges and Brahmaputra. The soil of the delta is amongst the most fertile in the world, replenished each year by the mighty flood waters of the monsoon season, heavy with nutritious sediment. The Jamuna and Meghna feed into the Bengal Basin from the north and east, while the Padma comes in from the north and west. The sources of the Ganges and the Brahmaputra are remarkably close, at the foot of Tibet's Crystal Mountain or Mount Kailish, holy to Hindus and Buddhists alike. These great rivers of India divide to drain land from Himalayan Nepal to the foothills of Assam, their waters pouring into the Bengal Basin where they unite again as tributaries of the Meghna.

Sand and gravel eroded by ice and water from the slopes of the still-growing Himalayas bring fertility to the Bengal plains. The steady grinding and weathering of rock releases precious nutrients. And arsenic.

o-o

The sediments of the Bengal Delta, also known as the Lower Ganges Plain, are geologically very young. About 17,000 years ago, when the sea level dropped by 130 metres during the last Ice Age, the Ganges and Brahmaputra rapidly cut deep channels into the land that rose out of the sea. As the sea level rose again, from about 14,000 years ago, this gouged-out plain quickly filled with the sediment carried to the delta by the rivers. Deposits 100 metres or so deep trapped water, forming aquifers. Into these aquifers aid agencies, governments and private individuals have sunk millions of tubewells.

When it dawned that the groundwaters of West Bengal and Bangladesh were poisoned scientists were confused. There is no precedent for aquifer systems like those in the Bengal Basin leaching arsenic. The geology of other known contaminated waters around the world, while not well understood, was, at least, very different from that of the Bengal Delta. Active volcanoes have seeped arsenic into groundwater in some parts of North and South America; elsewhere, such as the Altoplano Plain in South America, an extremely dry climate has concentrated arsenic and other toxic elements. Bangladesh and West Bengal have no active volcanoes, and the Bengal Basin is one of the wettest places on earth.

Only the slightest of clues could have forewarned of the disaster. Geologists found arsenic-tainted delta areas in south-west Taiwan in the 1960s, noting that they were very different from the other arsenic-rich zones of the world. The setting of Taiwan's aquifer is similar to that of the Bengal Basin, with one major exception. The arsenic in Taiwan is from bands of black shale: carbon-rich rocks like coal. There are no large-scale black shale deposits in the Lower Ganges plain, so no arsenic was expected. In hindsight, Taiwan's aquifers are all too similar to those of the Bengal Basin.

o-o

When arsenic was first properly reported in the groundwaters of West Bengal in 1983 local superstition decided that wells were being contaminated by venomous snakes living in the soil. Geologists and chemists came up with a more prosaic suggestion: that the arsenic was of human origin. During the first half of the 20th century the element was widely used in first-generation pesticides, such as the compounds lead arsenate, Paris green (copper acetoarsenite), MMA and cacodylic acid. This was before the prominence of organochlorine pesticides such as DDT, HCH and 2,4-D in the 1950s. Natural sources of phosphate fertiliser, such as rock phosphate (calcium phosphate), are also high in arsenic. Mining and the smelting of base metals often lead to high levels of localised arsenic pollution. The massive urban and industrial sprawl of Calcutta was another potential source – the city even had its own Paris green factory.

But the farmers of West Bengal were too poor and too rural to have much access to pesticides and chemical fertilisers. They were also too remote to be affected by the megalopolis of Calcutta. The source of the arsenic had to be natural.

The hunt was on for arsenic-rich strata in the Ganges that once flowed through West Bengal, providing the waterfront to the British Imperial city of Calcutta. The Ganges writhed away from West Bengal into modern-day Bangladesh in the 16th century, until the Indian government appropriated part of it again in the 1970s. It was in this ancient Ganges track, along the current course of the Hooghly River, that geologists and chemists set about searching for sources of arsenic-rich deposits. They started to publish their theories in the mid-1990s.

o-o

Speculation was the order of the day. The first guess was that the arsenic was derived from surface outcrops of the iron–sulphur mineral pyrite, in the Copper Belt of the Indian state of Bihar. Pyrite often contains lots of arsenic. Coal deposits in the Damador Valley, just north of the Copper belt, were also implicated. Suspicion also fell on the Gondwana coal deposits of the Rajmahal Traps, hills skirting the Ganges Plain in India on the northwest border of Bangladesh. These deposits held to 200,000 ppb arsenic. Also fingered were isolated patches of sulphides comprising up to 0.8% arsenic in the Himalayan foothills of Darjeeling. It was thought that upstream arsenic-rich mineral deposits might be mobilised by the Ganges and then deposited in the Bengal Basin during the yearly flooding.

Dipankar Chakraborti and others thought these pyrite and coal sources were responsible for the contamination in West Bengal, which they then believed to be a localised problem. Unfortunately this 'local source of contamination' theory held back the search for arsenic elsewhere in the Bengal Delta, losing valuable years in defining the full extent of the catastrophe.

o-o

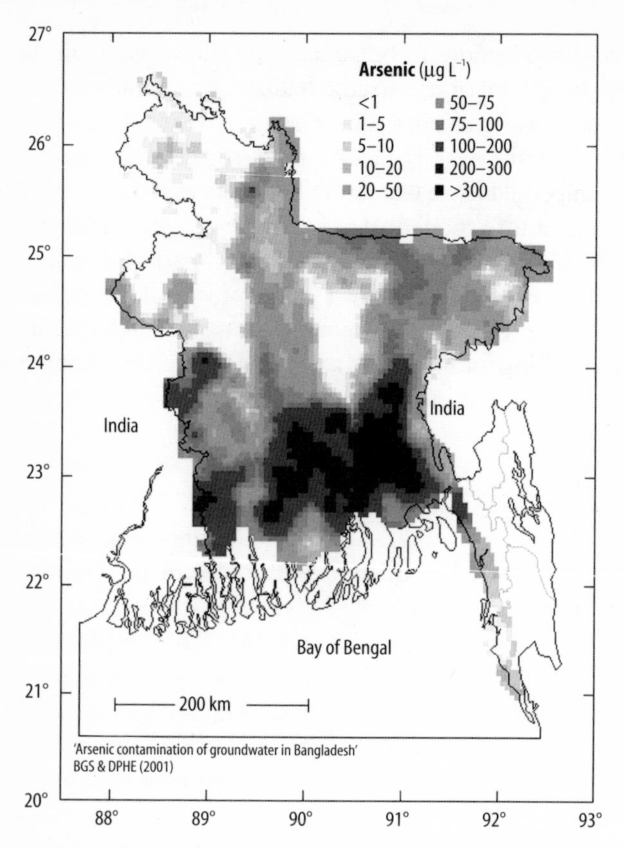

Figure 2.2 The British Geological Survey's arsenic map of Bangladesh. Reproduced with permission. Note: the unit 1 $\mu g\ L^{-1}$ is equivalent to 1 ppb, used throughout the book.

By 1999 the geographical distribution of the arsenic was mapped by the BGS (Figure 2.2). It curves from the northwest, where the Ganges enters Bangladesh, right up to the northeast corner of Bangladesh along the Meghna Valley. If the Ganges had contaminated the north-western and part of the southern tracts of Bangladesh, it certainly was not responsible for poisoning the Meghna. New sources for the abundant arsenic in vast areas of the Ganges–Brahmaputra–Meghna drainage basin had to be found.

The biggest mystery was that the rocks and sediment in affected aquifers have quite average arsenic levels, ranging from 1,000 to 30,000 ppb. Some of this arsenic is loosely bound to the surface of the sediments, while the bulk of it is encapsulated within, locked away from the surrounding water. Many rocks and sediments throughout the globe contain similar levels of arsenic without causing any build-up of arsenic in the water they hold. What is so different about the Bengal Basin? The answer seemed to lie in minerals that form part of the sediment. These pyrites and iron oxyhydroxides can contain very high levels of arsenic, and they gave rise to two separate theories.

Chakraborti and others proposed that pyrite was responsible for the contamination of West Bengal and Bangladesh. In 1996 his research group published their 'pyrite oxidation theory' in the journal *Environmental Geochemistry and Health*. They suggested that pyrite was either carried into the Bengal Basin from outcrops upstream or that it formed *in situ*. Under low oxygen conditions, certain bacteria convert sulphate, a common soluble ion in soil and groundwater, to insoluble sulphide. Sulphide reacts with iron, also common in soil and water, to form pyrite.

Elsewhere in the world, such as the Altoplano in South America, arsenic is released into groundwater when sulphide minerals react with oxygen; the insoluble sulphides become soluble sulphate again, and in the process release arsenic previously locked up in the mineral. Researchers in West Bengal understandably put two and two together. They postulated that pumping water out of the aquifers had lowered the water table, causing formerly oxygen-starved sediments to fill with oxygen.

The pyrite oxidation theory was questioned almost as soon as it was published. A 1998–1999 BGS survey of Bangladesh aquifers firmly discounted this theory, by showing that the waters were oxygen-starved, rather than oxygen rich. Conditions were *reducing* rather than *oxidising*, to use chemists' terms. Many presentations of this data were given at scientific meetings, but the urgency of setting up the large-scale surveys in Bangladesh meant that the findings only found their way into print later on.

The first published alternative to the pyrite theory, by Prosun Bhattacharya and his Swedish and West Bengali co-workers, came out in 1997 in the journal *Water Resources Development*. They raised the possibility that arsenic was released under reducing rather than oxidising conditions. The reducing theory was further elaborated in the journal *Nature*, in a 1998 article from geologist Ross Nickson and his colleagues from the Department of Geological Sciences at the University College London. They took water samples from tainted tubewells and looked for other elements. They found that, as arsenic levels increased, so too did quantities of bicarbonate and iron. High levels of iron and carbonate could only mean one thing: the bacteria underground were starved of oxygen, rather than basking in it.

Nickson's group suggested that the arsenic was locked away in iron oxyhydroxides. These chemicals are stable when there is plenty of oxygen, but when oxygen runs short, the minerals dissolve, freeing arsenic. Thus the same process that sets loose arsenic releases iron too – hence the presence of high levels of both iron and arsenic in tubewell waters. The bicarbonate comes from bacteria that digest organic matter. In using this energy source microbes also burn up any oxygen present, leading to oxygen-starved conditions. This sequence of events was termed the 'iron oxyhydroxide reducing' theory.

A major piece of evidence for the reducing theory is the form in which arsenic is pulled out of the aquifers. Arsenic in the wells is predominantly in inorganic forms. Under normal oxygen-rich conditions the oxidized form, arsenate (see Table 1.1, p. 10), dominates. When oxygen is scarce, arsenate is reduced to arsenite, which is unstable and which rapidly converts back to arsenate under oxygen-rich conditions. About half of the arsenic in tubewell water is in the form of arsenite, proof that reducing conditions persist.

The reducing theory has since become the dominant explanation, accepted by most scientists as the means by which arsenic is release into the aquifers below India and Bangladesh.

Nickson's paper resulted in three further publications in *Nature* in 1999, with the various proponents of opposing theories refining their stance. The BGS arsenic survey published in 2001 finally confirmed the pyrite oxidation theory. It presented overwhelming evidence for the iron oxyhydroxide reducing theory, showing the clear relationship between arsenic and iron levels in well water.

o–o

Only sediments laid down since the last Ice Age, and particularly those laid down in the last 7,000 years, release high levels of arsenic into the groundwater. Younger and older deposits – those shallower than 10 metres or deeper than 150 metres below the surface – release little of the element. Between these two levels lie the intermediate or Holocene sediments, laid down as sea level rose following the end of the last Ice Age. During this time the landscape was covered in mangrove swamp flooded when the tide came in.

Today's Bay of Bengal contains the largest mangrove swamp in the world: the Sundabans, meaning 'beautiful forest'. Wondering if this swamp held clues to the source of the arsenic, I visited a Sundabans forestry station. What I saw was sediment washed into the Bay of Bengal trapped around the mangrove roots. The sediments are packed so tightly that oxygen cannot get in. In response, the trees pump oxygen down and out of their roots to oxygenate their micro-environment. In this micro-environment, orange–brown bands of iron oxyhydroxides form readily, and oxidised arsenic binds easily to these minerals. Digging into the grey–green sediments reveals ochre streaks along the paths that the roots cut through the soil.

But there was another part to the arsenic-bearing sediment puzzle. Where was all the carbon coming from to create the reducing conditions? The answer to this question was also obvious when I visited a mangrove swamp. Stepping on the grey–green surface caused it to ooze a rich black slime that turned out to be chock-full of carbon and sulphur. The carbon simply comes from dead roots, leaves and bark shed into the forest floor. Elsewhere, over millions of years carbon in peat deposits turns to coal. In the relatively young Holocene sediments of the Bengal Basin, bacteria are still degrading the plant matter, consuming oxygen as they go.

Other clues lead back to mangrove swamps as the source of Bengal's woes. John McArthur from University College London and his co-workers have noted that peat beds are generally found in areas where groundwaters are high in arsenic. The locations of the peat beds basically show where Holocene mangrove swamps were. McArthur suggests that peat might be the source of carbon, fuelling the reducing conditions. But his theory has been criticised by some, as many of the arsenic-stuffed sediments are peat-free. I'd counter that: sediments can still be richly organic but not be peat. It is likely that organic sediments are found in and around the peat; peat only forms in specific patches of the mangrove swamp ecosystems.

Mangrove swamps are criss-crossed by thousands of rivulets that change greatly over time. Sediments laid down in these channels are lower in carbon than those formed under trees. The sediment layers, therefore, record an ever-changing sequence of surface landscapes. This could explain why two adjacent tubewells can have very different arsenic levels. A well that plumbs an ancient patch of trees will draw water heavy with the products of decaying organic matter. One nearby that draws on an ancient creek channel will not.

o–o

So a set of theories is in place to explain the accumulation and subsequent release of arsenic into the water. But the sediments have only average levels of arsenic compared to the rest of the earth's crust – between 1,000 and 30,000 ppb. How can they give rise to the 'greatest chemical disaster in human history'?

David Kinniburgh, a geochemist at of the BGS, has come up with a startlingly simple calculation to explain this conundrum. Knowing the amount of sediment and the amount of water in a unit volume of an aquifer, he calculates that if 1,000 ppb of arsenic dissolves out of the rock, this gives 6,000 ppb of arsenic in the groundwater. The highest observed values in the groundwaters of Bangladesh and West Bengal hover around 2,000 ppb arsenic. There is ample arsenic in the rocks to contaminate the water.

o-o

Many questions remain – not least why is it that delta aquifers high in arsenic have not been identified before? The answer seems to be that age decides whether or not an aquifer will release its toxic load. The water-bearing rocks that are leaching arsenic into their precious cargo formed during the Holocene. As sediments mature, their buried organic matter gets used up, and the liberation of arsenic abates. What's more, their mineralogy may change. Pyrite forms, locking up arsenic and, over time, iron oxyhydroxides may turn into more stable minerals such as haematite. In the West, where aquifer hydrology is most studied, Holocene aquifers are not as common or as extensive. In addition the tropical climate of south-east Asia may boost arsenic mobilisation – swamp forests produce a lot more carbon than the delta ecosystems of more northern latitudes.

One burning question remains: why is the problem much worse in West Bengal and Bangladesh at the moment than elsewhere? First, size: the Bengal Delta is the world's largest and most populous. A second factor is the advanced age of the tubewell program. finally, the high sediment accumulation rate in the Ganges Delta means there is a large depth of young material for tubewells to tap.

Chapter 3

FOOLS' GOLD

First he availed himself of poisons, those
He thought the surest and most virulent.
He found them all harmless and impotent.
'Vain help!' he cried. 'I carefully protected
Myself 'gainst each of them, and thus rejected
Unwisely the recourse I might have had
To them. Let us now turn to the surer aid
And seek a death more harmful unto Rome!'

Jean Racine, *Mithridate*, 1673. Translated by Lacy Lockert, Princeton University Press, 1958.

The earliest history of arsenic's relationship with mankind is recorded in something more durable than paper. It is cast in bronze.

Chemistry began as a practical business. Mineral, plant and animal extracts were pressed into use for tanning and dying, pigment and perfume making, and to produce glass, enamel and glazes. Advances were also made at the smith's forge.

From around 4500 BC in China and 3000 BC in the Middle East metallurgy was of prime economic and military importance, and smiths honed knowledge of how to develop the most suitable alloys for a given purpose. Bronze is an alloy of copper with either arsenic or tin. Most importantly, bronze alloys are harder than pure copper, so were used to produce weapons. Ores from different mines have different properties because of to the ratios of metals present. Adding certain minerals to molten metals changes their properties. Altering the heat or ventilation of a furnace affects its end product.

Between 3500 BC and 2500 BC arsenic-containing minerals were central to the development of bronzes. Minerals such as enargite (Cu_3AsS_3) are often found with copper ores. At first they would have been introduced by chance into copper smelting and refining, causing early metal workers to notice that ores from different mines made bronzes with varying hardness and strength. This would have led them to consciously select ores to achieve the properties they required.

In the 1960s archaeologists found a hoard of arsenical-bronze artefacts, including crowns, sceptres, standards and mace-heads, in Israel's Judean Desert. The haul dated from the early Bronze Age (c. 3150–2650 BC). Another such hoard, consisting mainly of weapons, was also found on the coastal plain of Israel. The nearest bronze-making sites to these hoards, in the Beer Sheva Valley, Israel, yielded a mace-head rich in arsenic. The closest copper mines to the hoards and the copper works, 150 km to the south, have ores low in arsenic, and could not have produced the alloys. The nearest arsenic-rich copper deposits are in the

Southern Sinai desert, approximately 350 kilometres away. The bronze workers of Beer Sheva, it seems, deliberately obtained arsenical ores from much further afield than their most local copper source.

Early Scottish and Irish Bronze Age weapons also illustrate the sophistication of the smiths. Rivets in bronze halberds (long-handled spear–axe hybrids) have lower levels of arsenic than the weapon bodies. Being softer than the axe-head, such rivets could be worked and closed without damaging the hilt. Analysis of the metals involved suggests that the smiths altered their forge conditions to pull off this particular trick, rather than selecting different ores for blades and rivets. Arsenic is relatively volatile compared with copper, and is readily driven off in the furnace.

From around 2500 BC, tin gradually replaced arsenic in bronze production. Why is still not clear – it was much scarcer and more expensive than arsenic minerals. The only known sources in Europe are a few outcrops on the remote Atlantic fringe. Either the ancients had tin sources that we still do not know about, or they imported all their tin from the Atlantic coast of Europe, the Far East and Africa.

The rise of tin–bronzes was probably driven by two factors. First, consistent results are hard to achieve in arsenic bronze-making. Second, the garlicky fumes of white arsenic trioxide given off during smelting are dangerous. Smiths would have been prone to lung cancer, nerve damage and muscle wasting. Of all the Greek deities described by Homer, only the god of fire and forge Hephaestus was deformed. Similarly, the Norse god of the forge Brokk was a dwarf. The smith's livelihood was not a healthy one.

o–o

The silver–grey element arsenic was named during the Renaissance from the Greek word *arsenikon*, meaning potent or male. *Arsenikon* was the name the Greeks gave to the poisonous yellow arsenic–sulphur mineral (chemical formula As_2S_3) that we now call by the Roman name orpiment, from *auri pigmentum* or gold pigment. However, *arsenikon* may have been derived from the Persian *az-zarnikh*, meaning yellow or gold. While the chemistry of the ancients was primitive, they seemed to know that orpiment was related to the red mineral AsS, which we call realgar from the name given by the Arabs, *rehj al-ghar*, meaning 'powder of the mine or cave'. Realgar was called sandarach by the ancients, from the root *sand-* or *sard-*, meaning red. Both the Greek physician Dioscurides (1st century AD) and the Roman natural historian Pliny the Elder (AD 23–79) wrote that these red and yellow minerals had a similar nature.

The Greek philosophers Plato (427–347 BC) and Aristotle (384–322 BC) believed that everything was composed of an indeterminate, mutable substance, called primordial matter, consisting of four elements: fire, air, water and earth. Acted on by heat, cold, moisture and dryness, this universal substance could be transmuted into all possible life forms and inorganic substances (Figure 3.1). Fire, for example, is the element formed when the qualities of hot and dry are impressed on the primordial matter. Water arises when cold and moist act on the primordial material.

This paradigm could be said to have kept most Western natural philosophers happy for the next two millennia.

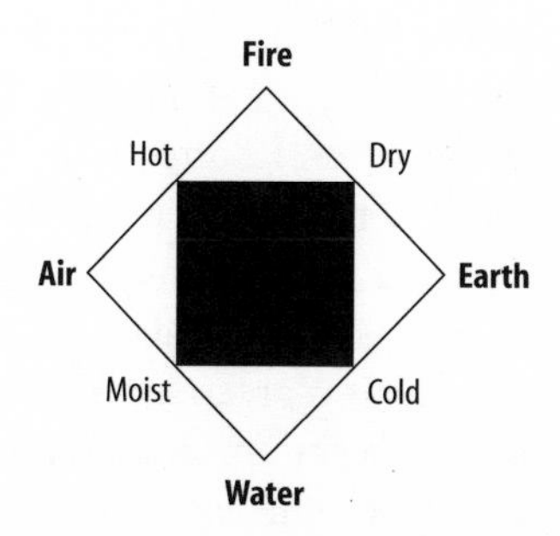

Figure 3.1 Aristotle's theory of everything.

This paradigm was applied to arsenic. Aristotle wrote in his *Meteorology, Book III*, that arsenic minerals were produced by exhalation (air) from the earth when heated by the sun: 'The ground gets hot and fire issues from it, and after this orpiment appears, descending from the impurity of the air, and fire, coupled with putrescence, is attracted (for this is orpiment)'.

His ideas were probably based on primary observations, or second-hand accounts, of the volcanic regions from which the Greeks would have obtained arsenic and other minerals. Realgar and orpiment form around fuming vents and hot springs.

In a world filled with natural hues, the harsh primary colours of yellow orpiment and red realgar were highly sought after. The Egyptians, Akkadians (Babylonians and Assyrians), Greeks and Romans used the minerals as paints and cosmetics. A linen bag of orpiment was found in Tutankhamun's tomb. Orpiment, probably from Persia, Armenia or Asia Minor, was used in wall paintings of the Theban necropolis, and extensively from the 18th dynasty onwards. Aristotle records that orpiment was used to colour hair red. Presumably it also got rid of head lice and dandruff.

o-o

Hippocrates was among the first to suggest using arsenic for medicinal purposes. Writing around 430 BC he gives a recipe for an ulcer lotion comprising realgar, orpiment, dried beetle, juniper oil, copper flakes, lead, sulphur and hellebore. Dioscurides noted that orpiment 'is astringent and corrosive.... It removes fungal flesh and is depilatory'. For realgar he notes 'mixed with pitch it serves to remove rough and deformed nails. With oil it is used to destroy lice, to resolve abscesses, to heal ulcers of the nose, mouth and fundament. With wine it is given for fetid expectoration and its vapours, with those of resin, are inhaled for chronic cough. With honey it clears the voice and with resin is good for shortness of breath'. Pliny the Elder gives a similar account. Galen (AD

130–200), the Roman physician and anatomist, recommended a paste of arsenic sulphide for ulcers.

o–o

The despot Nero was well acquainted with arsenic's poisonous properties. In AD 55 the Roman Emperor disposed of his rival to the throne, Britannicus (born ~AD 41), son of Emperor Claudius by his third wife Messallina. Nero was Claudius's adopted son by marriage to Agrippina. After Nero became emperor Agrippina encouraged him to poison Britannicus at a banquet. The Roman historian Suetonius vividly records this event in his book *The Twelve Caesars*: 'He tried to poison Britannicus.... The drug came from an expert poisoner named Locusta, and when its action was more sluggish than he expected – the effect was violently laxative – he called for her, complaining that she had given him medicine instead of poison, and flogged her with his own hands. Locusta explained that she had reduced the dose to make the crime less obvious. "So you think that I am afraid of Julian law?" he said. The he led Locusta into his bedroom and stood over her while she concocted the fastest-working poison she knew.... That night at dinner Nero had what remained poured into Britannicus' cup. Britannicus dropped dead at the first taste'. The suspicious Britannicus employed a taster to ensure he was not poisoned. Nero engineered that a drink given to Britannicus was too hot. Having passed the taster's safety check, the drink was cooled with chilled water. It was the untested cold water that Nero had laced with poison.

What evidence is there that Locusta used arsenic? White arsenic, the poisoner's preferred form, is slow acting and has a strong laxative effect. Indeed, arsenic poisoning often escaped notice because, in many ways, it resembles cholera and dysentery.

Suetonius also wrote of one of Rome's foremost enemies, Mithridates, king of Pontus (reigned 120–63 BC), but revealed little of the mythology that had grown around the king. It falls to

Pliny the Elder to explain. 'When the mighty King Mithridates had been defeated, Gnaeus Pompeius found in a notebook of his, written in his own hand, a prescription for an antidote: 2 dried walnuts, 2 figs and 20 leaves of rue were to be pounded together with the addition of a pinch of salt. Anyone taking this on an empty stomach would be immune to all poison for the whole day'. Pliny later tells us this about Mithridates: 'Alone and unaided, he devised a plan to drink poison every day after first taking remedies, in order that, by accustoming himself to the poison, he might become immune to it'. He was also said to have protected himself by taking first small doses of poisons and then steadily increasing the dose so that he could regularly ingest a deadly quantity with no ill effects. His feats are immortalised by the A.E. Housman poem *Terrence, this is stupid stuff*:

They put arsenic in his meat
And stared aghast to watch him eat;

According to Plutarch's *Fall of the Roman Empire*, Mithridates had a more practical use for poison 'He also gave each of his friends deadly poison to carry with them, so that no one need fall into the hands of the enemy against his will'. Plutarch also records that Mithridates had poisoned many people, including his own son, Ariarathes. Finally the king committed suicide after another son, Pharnaces, revolted against him.

The myth of Mithridates was popularised again in the 17th century AD by the French Playwright Jean Racine, who wrote a tragedy about the King of Pontus. Racine was suspected of the poisoning of his mistress and star actress the Marquise du Parc. A 14-year-old Mozart produced a musical score for a libretto based on Racine's play *Mitridate rè di Ponto*. The composer's rival, Salieri, when he was insane at the end of his life, said that he was responsible for Mozart's death by poisoning him. Peter Schaffer's play of Mozart's life, *Amadeus*, has Salieri using arsenic as his poison. The curse of Mithridates?

o-o

Practical technologies such as metalworking gave rise to tools that modern chemists would recognise: pestles, mortars, tongs and crucibles. But alongside such quotidian developments flourished chemistry's shady forerunner, alchemy.

If a smith could heat a mineral and make solid copper, with entirely different properties, surely rocks could be transmuted into something more valuable, namely gold? If all life was composed of one primordial material, which altered under the influence of heat, dryness, moisture and coldness, shouldn't it be possible somehow to create gold by treating a rock or a cheap metal to a combination of these different conditions? Gold was the most stable of the metals known to the ancients. According to Aristotle's theory of everything it contained the four elements – fire, earth, water and air – in perfect balance in an inseparable proportion.

Alchemists believed in a potent transmuting agent, the Philosopher's Stone, that would convert base materials into silver and gold. They believed that this miraculous stuff, if only they could find it, would also heal illness and prolong life; hence its other name: 'The Elixir of Life'.

At the time orpiment would have been an obvious candidate for transmuting into a precious metal. It is often found bound up with gold deposits, and is flaxen coloured. The Roman Emperor Diocletian (AD 260) was so upset with the failure of Egyptian alchemists to extract gold from orpiment that he had all the books dealing with transmutations destroyed.

Pliny the Elder writes: 'There is a method of making gold from orpiment, which is mined in Syria for painters; it is found on the surface and has the colour of gold, but is brittle like selenite. Its potential attracted the Emperor Gaius Caligula who was obsessed with gold. He ordered a great weight of orpiment to be melted; and certainly it produced excellent gold, but the yield was very low and so, although orpiment sold for 4 denarii a pound, he lost out by the experiment that his greed had led him to initiate. The experiment was not subsequently repeated by anyone else'.

o–o

The belief that base materials could be converted into riches led a wide range of substances to be subjected to a battery of chemical investigations. Alchemy therefore brought forth many discoveries in practical chemistry, despite its arcane and secretive nature. Practitioners often recorded details of their quest for gold or eternal life in code to stop their ideas from being stolen. It was not unknown for alchemists to be kept prisoner by patrons, often princes and monarchs, wanting to ensure that alchemical secrets did not fall into the hands of rivals. Texts attributed to renowned masters of the art were highly prized; those of unknown alchemists less so. Thus forgery and misattribution was rife, charlatans abounded and alchemy came to be considered a disreputable occupation.

Alchemists also developed their own techno-babble, making their writings impenetrable to non-initiates. While today's immunology or particle physics jargon is equally mystifying to the non-specialist, it is at least formalised and consistent. That certainly was not the case for the quixotic language of the alchemists, leading to much confusion.

Worse still, alchemical theories were embedded in mysticism, giving texts another layer of incomprehensibility. In ancient and medieval cultures numerology and astrology dominated mathematics and astronomy. So simple numeric ratios were believed to hold the secrets of the four Aristotelian elements that composed all matter, and the movement of the stars and planets were thought to influence the properties of that matter.

o–o

Just nine chemical elements were known to the ancient world: the metals gold, silver, copper, iron, lead, tin and mercury, and the non-metals sulphur and carbon. Arab and medieval alchemists added three more to this list: the soft metallic arsenic, antimony and bismuth. They felt sure that these three new elements were related to mercury, especially since realgar looks like the red powdery mercury ore cinnabar.

Following the dissolution of the Roman Empire around the 5th century AD, Europe fell into the cultural slumber of the Dark Ages. Luckily, another great civilization, the Arabs, salvaged the knowledge of Mediterranean Europe and greatly advanced on it, particularly in mathematics, medicine and chemistry. It was the Arabs who would hand back to 13th century Europe the scientific wisdom of their predecessors, through their translations of the Greek and Roman philosophers.

Nestorian Christians expelled from Constantinople in AD 431 moved to Edessa in northern Syria, taking with them what was left of Greek learning. The Nestorians travelled around Asia Minor ending up in Jundishapur in Persia. The Nestorians translated texts on medicine, religion, astronomy, mathematics and alchemy into Syriac Greek. The Arab world expanded rapidly after the death of Mohammed in AD 632, capturing Damascus, Jerusalem, Egypt, Palestine, Syria, Asia Minor, Crete, Sicily, Rhodes, Cyprus and North Africa, winding up in Spain in the 8th century. Nestorian Christians then started to translate texts into Arabic. This led to a fusion of Greek, Egypto-Babylonian and Muslim culture and a flowering of science, especially chemistry.

The most renowned of all the Arabian chemists was Abu Musa Jabir ibn Hayyan, known simply as Jabir, or Geber in Latin. There is some doubt as to whether he actually existed or not, but his story is given in *The Fihrist*, a bibliography of the Muslim world completed in AD 987. *The Fihrist* tells us that if Jabir was real, he was born between AD 720 and 723, flourished around AD 760 and wrote only one text, *The Book of Mercy*. The Shiite sect claimed him as a spiritual guide, some philosophers counted him among their number, and seekers of the Philosopher's Stone felt he was one of them.

Some 2000 Arabic alchemical works are attributed to Jabir. Still further texts were attributed to Geber up to the 13th century in Europe. Now only fragments of the so-called Jabir texts survive, ironically, as Latin translations; question marks hang over the authorship of all.

It is likely that Shiite mystics produced texts and attributed them to Jabir in the 9th and 10th centuries. The sect combined practical chemistry with mystical doctrines, and its members were very much disciples of Aristotle. If you disregard the mysticism and outmoded philosophy of the works attributed to Jabir, what is left are remarkable and advanced treatises on practical chemistry. The texts describe the earliest known preparation of nitric acid and the concentration of acetic acid through the distillation of vinegar. Since these acids dissolve minerals and metals, their manufacture led to other discoveries and preparations. Steel making and metal refining are described, linking the alchemists to the craft of the smiths. Dyeing, waterproofing and tanning are also outlined.

Amongst the numerous discoveries of the Arab chemists in works attributed to Jabir was the existence of arsenic in its metallic state. One book, translated into Latin as *De Fornacibus*, states that white arsenic (arsenic trioxide) contains a metal. Another, *Furnaces*, mentions metallic arsenic (*arsenicum metallinum*). Recipes are given for the production of arsenic trioxide by sublimation of orpiment. The book *Sum of Perfection* states 'Arsenic, which before its sublimation was evil and prone to adjustion, after its sublimation, suffers not itself to be inflamed; only resides without inflammation'.

Practical uses of arsenic are described in the 'Jabir' texts. Heating copper with arsenic, they report, turns it silvery – a transformation which would have piqued alchemical interest. The writings give a preparation of iron arsenate. 'Grind one pound of iron fillings with half a pound of sublimed arsenic [arsenic trioxide]. Imbibe the mixture with the water of saltpetre [sodium nitrate], and salt-alkali, repeating the imbibitions thrice. Then make it flow with violent fire, and you will have your iron white'.

o-o

The recapture of southern Spain by the Spanish during the 11th century brought the Christian West into contact with Arabic and Hebrew learning. The Archbishop of Toledo established a college that, on 11 February 1144, completed the first translation of an Arabic alchemical text into Latin: *The Book of the Composition of Alchemy*. This was supposedly authored by a Christian aesthetic, Morienus, who taught the secrets of alchemy to the Arab prince Khalid ibn Yazid.

One of the key instigators of the scientific and philosophical revival of Western Europe was Albertus Magnus (1193–1280; Figure 3.2). Born in Suabia, Magnus was supposedly an excessively stupid youth, but devoted to the Virgin. This devotion was rewarded with a vision accompanied by intellectual enlightenment. He became a Dominican monk, and was appointed to the Bishopric of Regensburg in 1260, which he subsequently resigned to pursue his scientific and philosophic studies in a Cologne retreat. In old age he relapsed into the mediocrity of his youth, giving rise to the saying 'from an ass he was transformed into a philosopher, and from a philosopher he returned into an ass'.

Albertus Magnus taught St Thomas Aquinas and led the revival of Aristotelian philosophy that dominated the European Renaissance. Magnus had a good working knowledge of practical chemistry – what equipment was required for distillation and sublimation and so on – and he believed in alchemy. Like other revered intellects of medieval Europe, many texts have been attributed to him, including the popular *Book of Secrets, of the virtues of herbs, stones and certain beasts, also a book on the marvels of the world*. Such encyclopaedias of fantastic facts were published up to the 17th century, satisfying readers hungry for weird and wonderful phenomena.

Magnus is also credited with a number of alchemical texts, one of which gives the method for the preparation of elemental

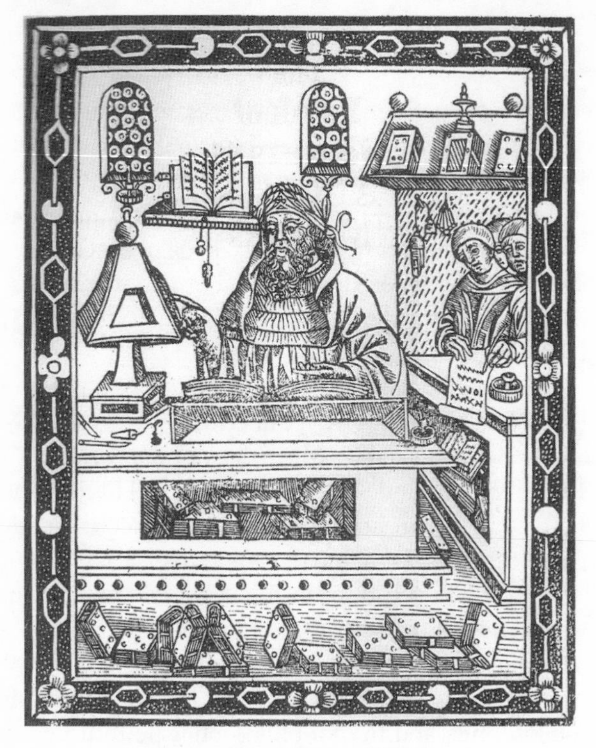

Figure 3.2 Albertus Magnus.

arsenic: 'if arsenicum [orpiment] is heated with twice its weight in soap it becomes metallic'. Whether this was a new discovery, re-discovery, or a quotation of a since-lost Arabic text, we can only speculate.

Facts gleaned from Arabic texts were often reported without attribution, or falsely credited to famed medieval alchemists. One such was Basil Valentine from Alsace, born in 1394. Biographical details in works supposedly authored by him suggest he was a widely travelled cleric of the Benedictine order. Valentine's documents reveal a good knowledge of arsenic and its compounds. Arsenic trioxide is called *hütten-rauch*, furnace smoke, and is 'a poisonous volatile bird' driven off during the smelting of ores. Arsenic is described as: 'white, yellow and red, and gives metallic arsenic (arsenicum) which is like mercury and antimony, but is useless in the transmutation of metals. It can be sublimed by itself and also with addition in various ways. When it is sublimed with salt or iron it is as transparent as crystal. By fusing white arsenic with saltpetre a deliquescent salt [potassium arsenate] is obtained'.

o-o

The first fictional depiction of an alchemist is in Chaucer's *Canterbury Tales*. The character appears in *The Canon Yeoman's Tale*, written in the 1390s, when alchemy in Europe was relatively new. Scholars believe that Chaucer drew on personal experiences and his reading of alchemical texts. Fiction's first alchemist is a conman given to simple tricks – such as roasting a mixture of silver and mercury to leave behind dissolved silver in the crucible. Chaucer appears to have a good chemical knowledge for his time; he knew, for example that orpiment and 'arsenyk' were key:

Oure orpyment and sublymed mercurie,
Oure grounden litarge [lead monoxide] *eek on the porfurie* [mortar],
Of ech of thise of ounces a certeyn –
Noght helpeth us, oure labour is in veyn.

The popular image of the alchemist did not improve much with the passing centuries. Ben Jonson's 1610 satire *The Alchemist* developed Chaucer's charlatan theme and arsenic was still part of the toolkit, along with one or two more easily obtained ingredients:

Your lato [brass], *azoch* [mercury], *zernich* [arsenic], *chirbrit* [sulphur], *heautarit* [mercury],
And then, your red man, and your white women,

With all your broths, your menstrues, and materials,
Of piss, and eggshells, women's terms, man's blood,
Hair o; the head, burnt clouts, chalk, merds [faeces], *and clay,*
Powder of bones, scalings of iron, glass,
And worlds of other strange ingredients,
Would burst a man to name?

Between Chaucer and Jonson seeing the preposterous nature of many charlatan alchemists, the revered scientist and practising alchemist Leonardo da Vinci (1452–1519) thought up a particularly unpleasant use of arsenic. Enlightened, gay, vegetarian and humanist, da Vinci was not above making money by designing weapons of mass destruction for his patrons, including the Renaissance ogre Cesare Borgia. He devised an early example of chemical warfare, an asphyxiating bomb comprising realgar, sulphur and bird feathers, which when lit would give off poisonous gas, *fumo mortale,* for seven or eight hours. It would be directed towards the enemy using smith's bellows. To keep the bellow operator safe from a change in wind direction Leonardo suggested an antidote for arsenic poisoning: rose water. At least the poor fellow would have smelt nice for his funeral.

o–o

In 1347 the Black Death arrived on the shores of Italy. Over the next four hundred years, wave after wave of the disease crippled Europe. Syphilis suddenly emerged in Europe around 1495 following Columbus' voyages to the New World. Medical practice in medieval Europe had no answer for either disease. At the same time, the Renaissance was bringing to light texts of the ancient Greek and Roman medics: Hippocrates, Galen, Celsus and others.

Medicine as a science essentially began with the Greek Hippocrates circa 430 BC. He realised that illness spread and progressed in a recognisable manner. His findings were unfortunately ignored by the Roman experimental physiologist Galen (2nd century AD) who constructed a new, inflexible creed, so self-contained that it quashed further medical research or original thought. Galen believed that ill health was an interaction between a patient's temperament, atmospheric conditions, and excessive eating or drinking. Galenic medicine was based on four humours – blood, phlegm, yellow bile and black bile – a further extension of Aristotle's element theory. A balance of these four led to health; imbalance resulted in sickness. Good health could be returned by a systematic, mathematical approach in which any case history had a logical and satisfactory explanation. Unfortunately, the logic, while self-consistent, was wrong. Nevertheless, this philosophy gave hope to a society in desperation and was grasped with fervour. The first edition of Galen in Greek appeared in 1525.

Galen had witnessed a Black Death epidemic and he believed that such diseases were due to a corruption of the atmosphere. (Plague is actually caused by bacteria transmitted by the fleas living on rats.) Following Galenic principles, medieval medics thought that the air needed to be purified. Dry and scented woods such as juniper, ash, laurel, cypress, vine or rosemary were used as fumigants. Scented plants were kept indoors and floors were washed in vinegar and rose water.

Aristotelian-inspired medical theories were as useful in curing disease as Aristotelian physics would be in putting a man on the moon. But ignorance was bliss, and Galenic medicine offered concepts that were understandable and holistic. However, no sooner had Galenic medicine become popular than its premises were challenged to the core.

o–o

Theophrastus Bombast von Höhenheim (AD 1493–1541) is probably the most notorious alchemist of all time. He was exuberant, charismatic, controversial, confrontational and, not least, boastful. He took the name Paracelsus, considering himself greater than the famed Roman physician Celsus. Paracelsus (Figure 3.3) urged alchemists to produce medicines rather than gold. He said that the whole world was a chemist's shop where cures might be found.

His life is a confused mixture of fact and legend. Born in Einsiedeln in Switzerland around 1493. His father was Wilhelm von Höhenheim, a medic who passed his knowledge on to his

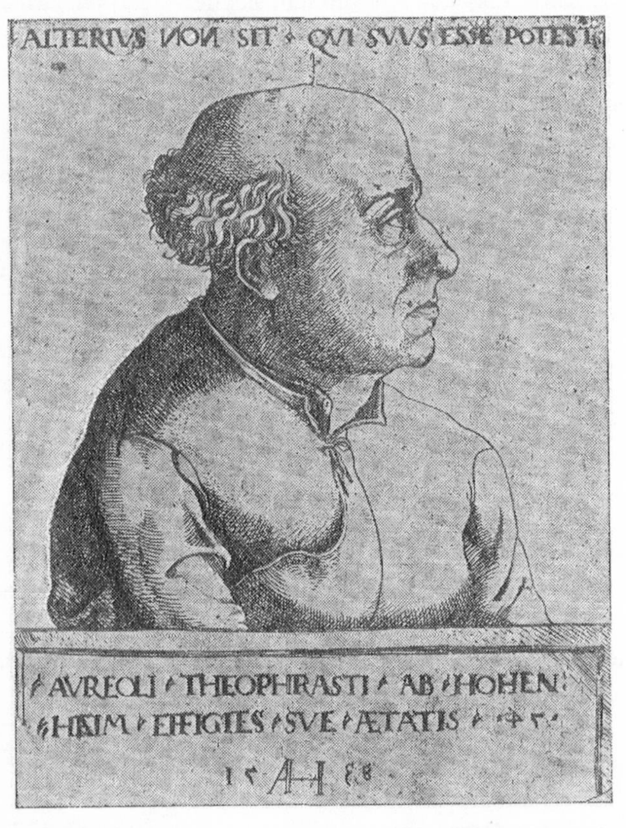

Figure 3.3 Paracelsus. A copy after Augustus Hirschvogel, dated 1538 and laterally reversed. Reproduced with permission from the Dept of Prints and Drawings of the Zentralbibliothek Zürich.

son, along with some chemistry and mineralogy. This teaching was reinforced by the metallurgy Paracelsus picked up from neighbouring mines when the family moved to Huttenburg, Austria, in 1502. By 1522 Paracelsus was allegedly an army surgeon.

Paracelsus hated qualified medical practitioners, possibly because he was a surgeon, a profession of lower standing. However, in 1527 he issued a statement claiming he had graduated as a medical doctor in Ferrara, although when he moved to Strasbourg in 1525 he was enrolled in the guild of grain merchants, not doctors. In any case, during March 1527, he was appointed officer of health in Basle, a municipal appointment that did not require a medical degree.

In Basle he made quite an impression. He began a lecture on 'The greatest secret in medicine' by uncovering a pan of excrement, causing his invited audience of physicians to march out in anger. On St John's day 1527 he denounced the antiquated systems of Galen, burning valuable books on traditional treatments, and calling instead for a new approach to medicine based on observation. Condemned as blasphemous, illiterate and fanatical, he left Basle in 1528 to escape impending arrest.

Paracelsus was a practical physician, with a respectable medical knowledge for his time and a reputation as a healer. Yet his works are imbued with the Christian and cabalistic mysticism of the apocalyptic theologians that flourished in 16th century Germany. He was very much a man of his time, and although he embodied a new way of thinking about medicine and chemistry, he could do so only in the language of the Renaissance. Similarly, while criticising Galenic medicine, making many enemies, he still basically used Galenic treatments. That said, his apparently nonsensical writings were well advanced over those of many of his contemporaries. The fog of alchemical illusion falls away once the goal is the preservation of human life, not the greed for gold.

Typically for his age, many texts were attributed to him, with 122 of chemical interest. Paracelsus's chemistry is mostly contained in nine genuine books called *Archidoxis*, written somewhere between

1525 and 1527 and first published in Cracow in 1569. Many of Paracelsus's chemical discoveries were simply raided, as was then the practice, from old texts. But his abilities as a self-publicist were unmatched. As more reputable German medical chemist Andear Libavius (c. 1540–1616) put it, rather touchily, 'chemistry was not invented by Paracelsus'.

⚯

Inevitably, one of the central elements in Paracelsus's pharmacopoeia was arsenic. After all, it was he who said: 'The only difference between a poison and a medicine is dose'. He believed that poisons were powerful remedies if deprived of their lethal properties by chemical treatment. A poison must be 'killed', 'sweetened' or 'fixed', as he put it. Paracelsus gives many recipes for preparing medicinal forms of arsenic, to treat, among other things, ulcers, wounds, gangrene and fistulas.

In a revealing passage from his work, titled *Concerning the true and perfect special Arcanum (incorporeal, immortal substance with its nature far above the understanding of man) of Arsenic for the White Tincture*, Paracelsus, relates arsenic to mercury and sulphur, describes how it can be used in copper transmutation, rages against Geber (Jabir), Albertus Magnus and Aristotle for giving wrong facts about it, and ends with the practical uses of arsenic in treating skin conditions:

> Some have written that arsenic is a compound of mercury and sulphur, others of earth and water, but most writers say that it is of the nature of sulphur. But, however that might be, its nature is such that it transmutes red copper into white. It may also be brought to such a perfect state of preparation as to be able to tinge. But this is not done in the way pointed out by such evil sophists as Geber in 'The sum of Perfection', Albertus Magnus, Aristotle the chemist in 'The book of the Perfect Magistery', Rasis and Polydorus; for

> those writers, however many they be, are themselves in error, or else they write falsely out of sheer envy, and put forth receipts whilst not ignorant of the truth. Arsenic contains within itself three natural spirits. The first is volatile, combustible, corrosive, and penetrating all metals. The second spirit is crystalline and sweet. The third is a tingeing spirit separated from the others before mentioned. True philosophers seek for these three natural properties in arsenic with a view to the perfect projection of the wise men. But those barbers who practice surgery seek after that sweet and crystalline nature separated from the tingeing spirit for use in the cure of wounds, buboes, carbuncles, anthrax and other similar ulcers which are not curable save by gentle means.

o-o

The work of Paracelsus pales besides that of the father of modern chemistry. Robert Boyle (Figure 3.4) was born to a privileged life in Lismore Castle, Co. Wexford, Ireland, in 1627, the seventh son and 14th or 15th child of Richard Boyle, first Earl of Cork. Educated at Eton, he became a Calvinist after a two-year sojourn in Geneva. He led a simple life, never marrying and turning down a peerage. He was among the first to pursue chemistry as worthy of study on its own, not merely as an aid to medicine or alchemy.

Boyle was in at the birth of European Enlightenment, when rigorous science started to blossom. He rejected the dubious theories of the past, and set out to construct new ones, laying down the principles of modern research practice. Boyle observed nature, made hypotheses and devised experiments to test them, and then refined or devised new hypotheses according to his results. Aristotelian philosophy just had the observation and theorising; experimentation was deemed unnecessary. Themistius, the Aristotelian in Boyle's epoch-changing commentary on the science of his time,

Figure 3.4 Robert Boyle.

The Sceptical Chymist, published in 1661, saw 'Paracelsus and some few other sooty empirics' of the previous century as the start of the anti-Aristotle revolution. Boyle recognized Paracelsus as an experimenter and a catalyst in the rejection of the philosophy that had stifled scientific theory since the time of the Greeks. Like da Vinci, Galileo, Kepler, Copernicus and Newton, Boyle started anew, opening up unexplored territory in our understanding of the mechanics of Nature.

Boyle lived before the great age of chemical discovery, when new elements were found and atomic theory developed. But he advanced many areas of science greatly: fire and flames, colour, luminescent compounds, acids and alkalis, the densities of metals and the composition of minerals. His name is immortalised in Boyle's Law, which describes the relationship between the volume of a gas and its pressure.

He discovered that in a vacuum, animals die, sound doesn't travel well, and smoke that would normally rise in air sinks. He published these findings in the first modern scientific thesis *Spring of the Air,* where experiments are described in detail without long philosophical discussions and references to accepted authorities. It was a clear exposition of careful conclusions based on observation, instead of being deduced by logic from accepted, but unproven, tenets of tradition.

o-o

Like Paracelsus before him, Boyle lived in a time that feared the Black Death. There was considerable debate and interest in its cause. Many thought that the plague resulted from substances in the air, with arsenic high on the list of culprits. In 1671 the Royal Society in London, of which Robert Boyle was a founding member, published a review of a book by Caroli de la Font, *The Nature and Cause of the Plague*, which implicated arsenic vapours emanating from the earth.

Boyle, typically, thought about the arsenic–plague link in a scientific way. He made careful notes on the effects of arsenic compounds on humans:

> The Observation is this; I knew, and on some occasions employ'd, a Chemical Laborant that fansi'd that he could make a rare Medicine out of red Arsenick, (as some call what others style Sandarach,) which is thought to differ little from common Orpiment, saving its been much higher colour'd. This Laborant then working long and assiduously upon this Mineral, and rubbing it frequently in

> a Mortar, came divers times to me; and complan'd of a disaffection he thence contracted in the Organs of Respiration; for which I gave him something that happen'd to relieve him; which encourag'd him to complain to me of another Distemper, that, though / not so dangerous, did often molest him; which was, that when he was very assiduous in the preparation of his Sandarach, it would give him great Pains, and (if I misremember not) some Tumours too, in his Testicles.

Then he noted that the tumours caused by arsenic exposure are similar to the buboes that plague gives rise to in the groin:

> What these mineral substances are, whose steams produce such odd and dismal Symtomes, I think exceedingly hard to determine. Yet, if I were to name one sort, I should perhaps think the least unlikely to be Orpiment. For, of the Poysonous Minerals we are acquainted with, I know not any of which there is greater quantities in the Bowels of the Earth; especially taking that name, in the latitude allowed it, by those skilful men, that make three sorts of it, vis. Yellow, Red and White Orpiment, divers of whose mischievous Effects seem to agree well enough with the Symtomes of some Plagues, / and may be guessed to have at least a considerable interest in the production of them.

But having conceded this link between arsenic and the plague, he concludes that he does not believe them to be connected, as arsenic does not result in the full suite of plague symptoms:

> ...to speak candidly, I do not think that these Minerals are the causes, even of all those Pestilences whose efficients may come from under the Ground... that Mineral Effluvia, may not only be noxious in a general way, but may produce this or that determinate Disease.

While arsenic was thought to be the cause of the plague, it was used to ward off the disease in a homeopathic, Paracelsian approach – treating *like with like*.

The earliest plague treatise in English – Old Scots to be precise – is from 1568, by Gilbert Skene, Professor of Medicine at Kings College, Aberdeen. Skene describes a plague outbreak in Edinburgh during which small cakes of arsenic were worn under the armpits where the buboes usually formed. Robert Boyle deduced the true impact of these poultices:

> That Arsenical Appensa, though much extoll'd by the divers Physicians themselves, and sold dear by Empyricks [followers of Paracelsus], as (if worn near the Heart) wonderful Amulets against the Plague, I have especially in some persons and circumstances) produc'd some of the noxious effects of Arsenical Poysons, and particularly caus'd in some great faintness and dispiritness.

He concludes that the only use of arsenic in fighting off the plague is in what we would today call its placebo effect:

> Drug is employ'd, as preservatives from the Plague, against which, I doubt the chief Succours they afford, proceeds from the Confidence or Fearlessness they give those that wear them.

While Boyle did not think that arsenic amulets were up to much, he still thought that arsenic was useful. He was interested in the idea that poisons could be made into medicines and described how arsenic salts 'maybe so prepared, as to become fit and safe to be taken inwardly'.

However, he later records a near-fatal arsenic ingestion experiment:

> I remember that a skilful Chymist, having in my presence tasted some prepar'd, and, as was thought, somewhat

corrected Arsenick; was quickly invaded by such Symtomes, as he thought would presently kill him. But, through God's blessing, I quickly put him out of danger, though not out of pain, by early prescribing him store of Oil of sweet Almonds, and something made of Lemons, that I chanc'd to have by me.

o-o

From Boyle onwards, chemistry developed rapidly and studies of arsenic played a big part in this advance.

The Swedish chemist George Brandt is credited with the earliest methodical investigation of arsenic substances. He made perhaps the most significant observation on this element in 1733, describing metallic arsenic as a *Halbmetall*: literally, a demi-metal. Today we have replaced the *demi-* with *semi-*, but Brandt's observation that arsenic has only some of the characteristics of metals matches how we classify the element. Arsenic, like its relatives antimony, tin, selenium, mercury, lead and bismuth, is grey and shiny. Unlike metals, it readily forms compounds with carbon, such as the highly poisonous arsenic gas trimethyl arsine (chemical formula: $As(CH_3)_3$). Indeed, the chemistry of arsenic resembles that of the non-metal phosphorus more than that of any other of the 100-plus elements. Arsenic lies below phosphorus, in the same group, in the Periodic Table.

In 1755 the Swedish chemist Karl Scheele, whose role in the history of arsenic is pivotal and to whom we will return, discovered the gas arsine (AsH_3). This is the most toxic form of arsenic, and many experimenters suffered dire consequences investigating it.

The last, fatal, study of German chemist Adolph Gehlen (1775–1815) was of arsine. Other chemists survived, including one of Enlightenment's greatest figures Joseph Priestley (1733–1804). Priestley was a great non-conformist thinker, Unitarian pastor and supporter of the French Revolution. He had to flee to revolutionary America because of his politics. Priestley was also an amateur chemist, credited with the discovery of oxygen, along with Scheele, who discovered it independently. The pastor describes smelling the 'strongly inflammable air' given off in an experiment where he passed steam over heated arsenic. It had a scent which 'could not be distinguished from that of phosphorus'. It would certainly have contained arsine.

o-o

Descriptions of the most remarkable, and dangerous, of arsenic synthesis experiments were published in 1760 by a French pharmacist, Louis-Claude Cadet de Gassicourt (1731–1799), known simply as Cadet. He describes carefully distilling powdered arsenic trioxide with dry potassium acetate in an attempt to make elemental arsenic. Instead, he produced an extraordinary, heavy brown fuming liquid with a very unpleasant smell that caught fire spontaneously on exposure to the air.

> If these liquids are exposed to air they fume like phosphorus and generate a very strong odour of garlic. These vapours do not inflame when exposed to a burning candle but on opening the grease-sealed receiver containing the two liquids to the air the caused immediate inflammation of the grease, which surprised me very much.

This stuff became known as Cadet's fuming liquor or alcarsin. A study of it by the German chemist Robert Wilhelm Bunsen more than 70 years later laid the foundation of modern organic chemistry. Compounds isolated from this liquid were given the name kakodyl (now cacodyl), from the Greek for 'stinking'. When cacodyl is oxidised by contact with water, cacodylic acid is formed (chemical formula: $As(CH_3)_2OOH$). This is much less toxic than inorganic arsenic compounds and became a widely used pesticide.

The preparation of cacodyl threatened the lives of the chemists who worked with it. Bunsen had to breathe through long glass tubes connected to the outdoors, and did some experiments in the open air. He lost some sight in one eye when a drop of cacodyl oxide fell on his hot distilling apparatus, causing an explosion, and 'lay for days between life and death'.

Chapter 4
THE VERDANT ASSASSIN

Green arsenic smeared on an egg-white cloth,
Crushed strawberries! Come, let us feast our eyes.

Ezra Pound, *L'Art*, 1910, published in *Personae* in 1926

Figure 4.1 Karl Scheele.

The synthesis of the arsenic salt copper arsenite ($CuAsHO_3$), published in the journal of the Stockholm Academy in 1778, led to the mass poisoning of the Victorian world via the unlikely media of wallpapers, sofas, stockings and blancmanges. Copper arsenite provided paint and pigment manufacturers with the vibrant green they had longed for.

Karl Scheele (Figure 4.1) was born in Pomerania in 1742. At fourteen he was apprenticed to an apothecary in Gothenburg for eight years. In 1765 he went to Malmö, in 1767 Stockholm, and in 1773 he moved to Uppsala to become an apothecary's assistant. By 1775 he ran a business in Köping, which he bought two years later. In 1786 he married the widow of the previous owner and within 48 hours he was dead.

In his spare time Scheele revolutionised chemistry. He discovered chlorine, oxygen, ammonia and hydrochloric acid gas, a range of acids and a host of organic chemicals.

In 1775 Scheele discovered arsenic acid (H_3AsO_4) by reacting nitric acid (HNO_3) with white arsenic (As_2O_3). Then, in 1778, he found that one of the salts of the reduced form of arsenic acid, arsenious acid, included 'a new green colour', copper arsenite ($CuAs_2O_3$), produced by mixing arsenious acid with ammonium sulphate of copper.

The artist Joseph Turner used Scheele's green in 1805 in his oil sketch of Guildford from the banks of the Wey and Edouard Manet used it in his *Music in the Tuileries Gardens* (1862). Much more popular with painters was the colour Emerald green. This copper aceto-arsenite ($3CuO.As_2O_3.Cu[OOC.CH_3]$) was discovered by the German paint manufacturer Wilhelm Sattler and was made by the reaction of vinegar and white arsenic on verdigris (copper carbonate). It was not widely available till 1822, when the German chemist Justus von Liebig published a report of its synthesis, which Sattler had guarded closely. Windsor & Newton started selling it as an artist's oil paint in 1832, and its earliest use was by Turner from about this date. This brilliant green was a favourite with the Pre-Raphaelites and Impressionists.

Before Scheele's green and other arsenic–copper salts, green pigments were expensive and of limited hues. The most widely used was the expensive pale blue–green copper carbonate salt *verdigris*. Verdigris was prepared by exposing sheets of copper to vapours from wine vinegar. Another copper carbonate pigment, Green verditer, was produced by mixing copper sulphate with potassium carbonate. Natural minerals like malachite and glauconite were ground into dull pigments with a bluish tinge.

In 1764 pigment makers known as the brothers Gravenhorst of Brunswick introduced a pale green copper chloride salt called

Brunswick green that was embraced by oil painters but not interior decorators. The only green animal or plant pigment was sap green. This was prepared from the juice of ripe buckthorn berries and its main uses were the painting of ladies' fans and watercolours. Green pigments could be made by mixing yellow and blue components, such as the toxic concoction of chrome yellow and Prussian blue (iron ferrocyanide) named Prussian green. But artists or manufacturers did not favour these unnatural colours.

Scheele's green and Emerald green supplanted all others. The industrial revolution ensured that these poisons were carried into virtually every home.

⚯

Besides being vivid, arsenic greens were cheap. Arsenic was a by-product of the mining and smelting of metals such as copper, cobalt and tin. There was little commercial use for it, apart from rat poison, until the synthesis of copper arsenite and copper aceto-arsenite. On 4 September 1863 *The Times* reported:

> For some years past it has been a common practice... to employ for various ornamental and decorative purposes a peculiarly vivid green, produced by the use of arsenite of copper. The colour so obtained is not a natural one. It is not the green either of trees or plants or grass; but it is exceedingly brilliant and attractive...

With the onset of mass production, man-made pigments were needed to meet an ever-growing market clamouring for new and exciting products. The mills of northern England were churning out cotton goods at an unprecedented rate. Interior decoration was becoming fashionable and affordable. As early as 1800, large quantities of Scheele's green were used in wallpaper production.

Several innovations and tax concessions made wallpapers available to the masses. The Frenchman Nicholas-Louis Robert

(1761–1828) had invented a process for making continuous rolls of wallpaper in 1798, but their use and manufacture was banned by British Excise authorities until 1830. The 1836 budget removed all duty on stained paper, reducing the price of wallpaper further. Machine-printed papers came on the market in 1841, driving the price still lower. Paper duty was abolished on wallpapers in 1861, causing another large drop. Finally, in the 1860s wallpaper producers began to use wood pulp imported from North America and Scandinavia instead of the traditional recycled rags, of which there was a limited supply.

Driving the trend for wallpapers was the revolution in domestic lighting. Gas and oil lamps were replacing candles. As homes got lighter, the need for reflective wall coverings decreased, and wallpapers darkened. Deep reds and greens came to dominate the Victorian interior, the green supplied by copper arsenic salts.

As the *British Medical Journal* put it in 1871: 'in the majority of dwelling houses, from palace down to the navy's hut. It is rare to meet with a house where arsenic is not visible on the walls of at least some of the rooms'. Evidence of this chilling trend is still emerging. In 2004 I discovered a room so-papered in a humble farmhouse in one of the most remote parts of Britain – the National Trust for Scotland's Morlanich longhouse in the Scottish Highlands.

And papers exported from Britain contaminated homes in far-flung corners of the globe, including the mountainous tracts of India, according to an 1872 *British Medical Journal* article:

> A friend not long returned from India informs me that green papers are now frequently used in the Indian bungalows at the hills... persons who are sent to the hills for health, often inhale this poison [arsenic] unawares, and then wonder why the mountain air has done so little for them.

The amount of arsenic used in producing wall coverings was enormous. Papers commonly contained 25–35 grams of it per square metre. A large room requiring 100 square metres of paper might contain 2.5 kg of arsenic. It has been estimated that by 1858, some 260 million square kilometres of arsenic-containing papers were hanging in British homes.

o–o

Not long after the introduction of arsenic green wallpapers the deaths and illnesses started, usually in children. Throughout the second half of the nineteenth century, Britain's leading medical periodicals, *The Lancet* and the *British Medical Journal*, along with *The Times* newspaper, campaigned to rid society of the menace.

The Lancet's founder, Thomas Wakley, is thought to have survived poisoning by green wallpaper following the redecoration of the journal's Strand office. One of the journal's reporters, Robert Brudenell Carter, had even more cause to campaign. He wrote: 'two children of mine died... from the arsenical wall-paper in the nursery'.

In 1860 *The Lancet* asked: 'Are papermakers really unconscious of the dangers to which they expose their customers while luring them to buy green wallpapers? – would they be amenable to the charge of manslaughter... But can they be ignorant? It seems almost incredible'. During 1862 and 1863 *The Times* ran report after report of children killed by vapours given off by their bedroom walls. The editorial of 4 September 1863 mentioned above went on: 'Ever since the introduction of this new colour rumours from time to time reached the public ear of sufferings experienced by those who were engaged in its application, or mischief occasioned by the use of articles to which it had been applied. It is only lately, however, that a competent authority has pronounced upon its nature and effects, and that its injurious character has been clearly demonstrated'.

One G.J. Wigner, writing to the editor of the *The Lancet*, proclaimed in 1878: 'It seems to me that the colouring of paper with poisonous colours should be made a criminal offence'.

o-o

Arsenic came in other hues besides green. Red and yellow realgar and orpiment were also widely used in wallpapers. A whole host of aniline dye colours were often contaminated with arsenious acid used in the dyeing process. Sixty out of 70 wallpapers tested for *The Lancet* in 1877 contained arsenic, including those coloured blue, red, brown and pink.

By the mid-19th century, there was no reason why green pigments should have been made from arsenic, or indeed any other toxic element. A report in the *Transactions of the Royal Scottish Society of Arts* in 1883 found that a wide range of pigments were being mass-produced, all free of arsenic.

Not till the 1890s did arsenic greens stop being used in wallpapers, primarily due to the public protesting with their wallets, rather than government intervention. As late as 1931 two children were killed in the Forest of Dean in England by arsenic-containing wallpaper that had presumably been hung during the 19th century.

o-o

Originally, arsenic-containing dust shed by cheaply printed flock papers was thought to be causing deaths and illness. Traces of arsenic could have been inhaled, but the particles would have had to be very small to have made it into the lungs. The verdant assassin was actually much more insidious: it was the organic gas trimethyl arsine.

As early as 1815, several cases of arsenic poisoning in Germany had been ascribed to wallpaper coloured with green pigments. Leopold Gmelin (1788–1853), a distinguished chemist responsible for the systemisation of organic chemistry, worked out what was going on and published his findings in 1839 in a Sunday supplement of the newspaper *Karlsruher Zeitung* (the Government Gazette of the Grand Duchy of Baden). In 2002, the translation was published in an article by Thomas Chasteen, Markus Wiggli and Ronald Bentley in *Applied Organometallic Chemistry*.

Warning of certain green wallpapers and paints

In newer times, for green wallpapers and room paintings, there is usually used a colour material, which is named Green of Schweinsfurt, of Vienna etc., and which impresses by the vividness of its colour, but which threatens health due to its considerable content of arsenic. Only in very dry rooms is there nothing to be afraid of, especially those facing to south and which are not on the ground floor and are regularly heated and aired. On the other hand, in rooms facing to the north and being on the ground floor, and in those which are not heated itself, but where the warm vapour of the adjoining room can enter, the moisture settling on the walls causes a slow decomposition process of the paper and the paste, in which the green colour is dragged in. The result of this is a mouse-like odour, which is easily noticed by entering a room not aired for some time. There is no doubt that this smell is caused by a trace of arsenic, which volatilises as a special compound (probably alcorsine). Brief inhalation of such is without danger, but longer daily stays in such rooms can cause harm, headache and undefined indisposition were noticed as a consequence of it, but even longer effect of this poisonous atmosphere can cause chronic poisoning by arsenic. Rooms which do not produce this bad odour, can be inhabited without concern. This smell can also stem from wallpapers of other colour, if there are only some green spots. Some wallpapers take years after pasting until the odour appears; it should not be hoped that it disappears again after a while; it can increase and decrease due to the moisture of the walls and the temperature, but most likely will only stop when the green colour is destroyed. To get rid of the bad odour and the danger of poisoning, it is necessary to carefully

> tear off the wallpaper; to over paste it would not help at all. These are the experiences which I have made over several years, especially numerous during the autumn, in this town and about which I feel obliged to publish. The question obtrudes if this colour material should not be prohibited totally for wallpapers and paintings, except in oils.

The explanation of arsenic dust causing the poisoning was discounted by Gmelin, as he noticed that poisonings occurred in rooms where the arsenical wallpaper had been covered with fresh arsenic-free paper. It is a pity that the British public did not pick up on this piece of detective work till much later.

In 1891 the Italian chemist Bartolomeo Gosio (1863–1944) isolated several fungi that could liberate arsenic gas. He noticed that a potato mash containing arsenic trioxide exposed to the microbes present in air gave off a garlic odour, not Gmelin's 'mouse-like' odour. The gas was named in his honour – Gosio's Gas – but Gosio did not work out its chemical structure. The gas was only properly identified as trimethyl arsine in the 1930s by the Yorkshire chemist Frederick Challenger. Other volatile substances produced by cheap wallpapers, such as sulphur and selenium, could have formed Gmelin's 'mouse-like' cocktail – or maybe Gmelin's sense of smell was wanting.

It was not just wallpaper pigments that caused trimethyl arsine production. Rat poison containing arsenic was mixed into wallpaper paste to deter rodents. This too reacted with damp to produce the gas.

The application of arsenical wallpapers was promoted as late as 1912. The book *The Expert Paper Hanger* states: 'It is possible, although improbable, that arsenic or its modifications may enter into the composition of wall-papers, although greens are least likely to contain them. Upon some reds suspicion might justly fall'. But advances in the understanding of arsenic chemistry had not reached this author. 'When one considers, furthermore, the very high temperature necessary to the volatilisation of arsenic, it is equally difficult to conceive of it entering the air at ordinary room temperatures'.

o-o

Poisoning by arsenic green wallpapers became so ingrained in the cultural consciousness of 19th century Britain that deadly green rooms entered into crime fiction. The popular arts and science magazine *Chambers Journal* published a tale in 1862 where an Emerald green bedroom was central to the plot. The story involves a rich orphaned boy, Sir Frederick Staunton, lodged with the poor local curate by the boy's evil grasping uncle Mr Richard Staunton. Mr Staunton is very specific about which bedroom the boy is to be housed in. The boy falls ill after a while in the curate's house. The local doctor cannot get to the bottom of the illness and the curate's wife Clara sends a telegram to a foremost London medic to come down and attend the boy.

> ...said the doctor; 'but you must move him at once. Any other room will do; but no time is to be lost. I have found out the real phantom-monk, the true destroyer that haunts your best bedroom.'
>
> 'What?'
>
> 'Arsenic!' said the doctor, exhibiting some powdered matter of various shades and tints, from dark green to pure white – 'arsenic enough to poison a regiment. That rich emerald green paper on your walls is stained by its means, and contains poison enough to be the death of generation after generation. I misdoubted it from the first. It has given me a headache, and is no doubt the cause of Sir Frederick's strange symptoms, and of the many untimely deaths that fatal room has witnessed...

Richard Staunton, being local gentry, knew the bedroom was a killer. The plot also mentions that Richard was interested in the chemical arts, implying that he knew why the room caused death, whilst other locals failed to make the connection.

o-o

Arsenic greens were a hazard beyond the home. Pigment manufacturing plants and printing rooms were extremely unhealthy places. In 1832 one C.T. Thackrah described illness amongst wallpaper stainers in his book on occupational health, *The Effects of Arts, Trades, etc. on the Health*:

> Paper stainers suffer chiefly from the rubbing and grinding of paint. When arsenic or white lead is employed they loose appetite and are affected with severe head-ache. Sickness often results from Prussian blue and arsenic, especially when turpentine is employed.

J.T. Arlidge, in *The Hygiene Diseases and Mortality of Occupations*, published in 1892, gave an assessment of the working environment of wallpaper factories:

> The tinting and staining of paper to be used for wrapping and ornamenting fancy goods, and also for wall-paper, is another industry in which arsenical colours are largely employed, and not only the greens, but some of the buff pigments. The work soon affects the artisans, who cannot follow it up for any length of time together.
>
> ...The mischievous consequences do not end with those concerned in manufacturing the paper, but extend to paper-hangers and to labourers engaged in stripping walls from their poisonous covering, and eventually to the occupants of the papered rooms.... These workers had arsenical conjunctivitis, discharge and ulceration of mucus membranes, fever, poor digestion, loss of appetite, irritable bowels, white tongue.

Later in the same book Arlidge writes:

> The ill consequences occur in the operation of drying, and in the carting and packing of the material [arsenic greens], and are exhibited by the development of boils and pimples, and by an itching rash about the nostrils and in the flexures of the arms. In severe cases headache, thirst and nausea are set up, together with an irritating eruption on the scrotum.

But the worst afflicted were those who made artificial flowers and leaves, tinted with Scheele's or Emerald green, for ladies bonnets and head-dresses. When these accessories became popular in the late 1850s, two to three hundred people, mainly girls, were employed to shape and colour the faux foliage by hand. Workers suffered myriad consequences, as *The Lancet* noted in 1861:

> The habitual empoisoning of young persons engaged in flower-making by inhalation of the powder of arsenite and copper used for colouring the green leaves and buds, has been the subject of frequent investigations and of open warnings in these pages.... In making artificial flowers, the most objectionable process is that known as 'fluffing', which consists in dipping the leaf into warm wax, then powdering it with Scheele's green by means of a dredger. Of necessity, such a process diffuses dust upon the workwoman, upon her clothing and hair, and surrounding objects. Other portions find their way under the nails and in the furrows of the skin; and in this way reach the mouth. Thus it happens that, by inhalation into the lungs, and by introduction within the alimentary canal, slow arsenical poisoning ensues, which is presently exhibited by chronic inflammation of the stomach and bowels, irritation of the eyes and skin, often also with accompanying eruption, great nervous debility, prostration, wasting, and the consequences of the gastro-intestinal lesion.

There were fatalities from this poisonous craft. *The Times* editorial of 1863 outlined in detail the suffering of artificial flower makers: 'It is in the manufacture of artificial flowers, fruits and leaves that the deleterious effects of this emerald green are most forcefully experienced... it has, in one instance, at least, probably in more than one, led to fatal consequences'. This death was of one Matilda Scheurer. The coroner returned a verdict of 'Death by arsenite of copper'.

o-o

Wallpapers are ephemeral and do not survive well. Those that have been preserved tend to be the most exquisite. These allow us to examine the use of arsenic greens in the 19th century and to understand the cultural context in which arsenic was invited into the home.

If asked to name a wallpaper designer, William Morris (1834–1896; Figure 4.2) would be the first name on most people's lips. Morris was a furious ball of creative energy and one of the greatest men of the Victorian era. He was a poet, novelist, calligrapher, medievalist, and a campaigner for the preservation of ancient buildings. He started the Arts and Crafts movement, and was one of the most influential interior designers of any age, setting up his famed company Morris, Marshall, Faulkner & Co. in 1861, renamed simply Morris & Co. when he acrimoniously disposed of the other partners in 1875. Whilst doing all of this he became very active in politics, developing into Britain's most famous Marxist. As a campaigner against industrial pollution he was the progenitor of the 20th century Green movement. In short, he was quite a man. For a Marxist and environmental activist though, he had a remarkable association with arsenic.

William Morris saw wallpapers as debased art, preferring tapestry hangings, hand-painted murals, embossed plaster, wood panelling or 'honest whitewash instead, on which sun and shadow

Figure 4.2 William Morris, Painted by W.B. Richmond in the 1880s. National Portrait Gallery, London. Reproduced with permission.

play so pleasantly'. He accepted that not all his clientèle could afford handcrafted interiors; his lush wallpapers found a ready market and became a mainstay of Morris & Co. In his own homes Morris tended to use more expensive approaches to wall decoration, although he did have some rooms fitted out with his own papers, most notably his bedroom at Kelmscott Manor which was lined with *Trellis*, his first wallpaper pattern. It was designed in 1862, and was the third of his papers to be printed, in 1864.

The *Trellis* pattern is an old-fashioned rose intertwined with geometrically repeating wooden squares, interspersed with birds. These birds haunted the dreams of one child, as described by May Morris, William's daughter. 'In my Father's Trellis there was a certain one of the birds who gave anxiety to a child in her cot high up in the Queen Square house because he was thought to be wicked and very alive'.

I was lent a scrap of *Trellis* (Figure 4.3) by the William Morris Gallery, Walthamstow. It was rescued from the house of William's foreman, George Campfield. The foliage was printed in arsenic green. Had the parents of the child known that the pigments used to create *Trellis* were highly poisonous, with the potential to release toxic arsenic vapours into their darling's bedroom, they too would have been lost in anxiety.

The wallpaper manufacturer Jeffrey & Co. was subcontracted to print Morris & Co. wallpapers. Jeffrey & Co. kept chronological records of the colours it used in printing Morris's papers in a logbook along with actual wallpaper samples. Morris personally approved the quality and colour of each sample in the book. This beautiful record of the colourways for each design is now in the

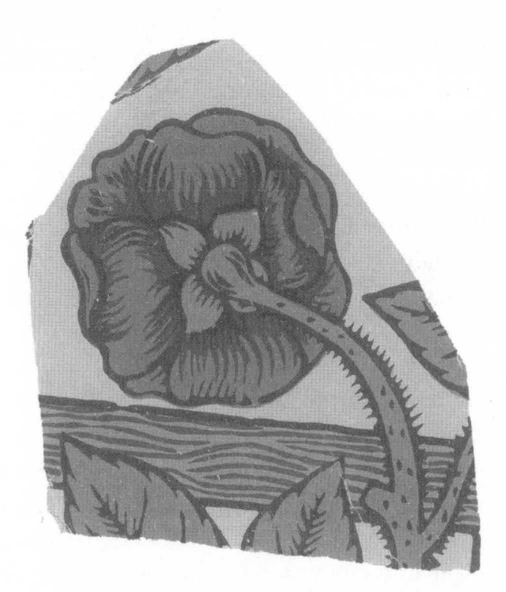

Figure 4.3 Scrap of William Morris' *Trellis* pattern wallpaper from the house of George Campfield, owned by the William Morris Gallery, Walthamstow, London.

possession of the interior design company Sanderson, which bought Morris & Co. in 1927 when it went into liquidation. Sanderson still prints from original Morris & Co. hand-carved blocks to this day. Managing director Michael Parry kindly let me analyse the logbook for arsenic using a non-destructive X-ray fluorescence machine. This looks for arsenic and other elements by recording the X-rays emitted by the pigments when X-rays from the instrument are shone on them.

Of the first eleven papers that Morris designed, nine contained arsenic: *Daisy*, *Fruit*, *Trellis*, *Venetian*, *Indian*, *Diaper*, *Spray*, *Scroll* and *Larkspur*. The two arsenic-free designs were a dark green version of *Queen Anne*, the seventh design, and *Branch*, the tenth design. *Branch* is actually an 'under print' of *Scroll*. *Scroll* has arsenic green flowers printed over arsenic-free branches. The very last design with arsenic was a later printing of *Indian* in green, which appeared in 1872 and is reproduced on the cover of this book.

Trellis, along with the equally dangerous *Daisy* and *Pomegranate*, were used in the 1867 decoration of Speke Hall, the home of the Liverpool shipping magnate Frederick Leyland. The artist Sir Edward Burne-Jones, a partner in Morris, Marshall, Faulkner & Co., had his house in Fulham kitted out with the firm's wallpapers. Little did this exclusive clientèle realise just how high a price they were paying for such fashionable fittings.

o-o

The wallpaper trade that made Morris & Co. famous, besides its use of poisonous pigments, had other very unpleasant aspects. As Morris leafed through his beloved copy of *Das Kapital* (which he read in French in 1882), did he pause? Marx discusses the trade's atrocious working hours and child labour. He writes of children toiling in wallpaper factories for $78\frac{1}{2}$ hours per week, quoting 13-year-old J. Lightbourne 'We worked last winter till 9 (evening), and the winter before till 10. I used to cry with sore feet every night last winter'. Marx also quotes distraught parent G. Aspen: 'That boy of mine.... When he was 7 years old I used to carry him on my back to and fro through the snow, and he used to have 16 hours a day'. There is no record that the working conditions at Jeffrey & Co. were any different from the rest of the wallpaper business. Marx does not comment on the toxic environment in which these children worked.

Morris's biographers have overlooked the fact that he used arsenic in his wallpapers. He is remembered for championing natural pigments following the advent of synthetic aniline dyes invented by William Perkin in the late 1850s. Synthetics were cheap, quick and easy to use and provided a wide range of colours, and the old methods disappeared quickly. Morris thought synthetic dyes were weak, garish and faded too fast. The following is from his essay *The Art of Dyeing*:

> Of these dyes it must be enough to say that their discovery, while conferring the greatest honour on the abstract science of chemistry, and while doing a great service to capitalists in their hunt after profits, has terribly injured the art of dyeing, and for the general public has nearly destroyed it as an art.

It was only later in his career, however, that Morris controlled what coloured his products. In the early days the companies to which Morris subcontracted used synthetic dyes.

Somewhere along the way, Morris's use of synthetics, such as Scheele's green, seems to have got lost in the myth. In the words of adoring historians such as Alan V. Sugden and John L. Edmundson, writing in *A History of English Wallpaper* (1926):

> 'No wonder he revelled in colour, as bright and full and as pure and clear as he could get it. All his colours, however, had to be vegetable colours, however, he was 'down' on synthetic colours, none of your 'chemical' dyes for him, though possibly he would have overcome his antipathy to them in

these days, when the range of colours includes every shade and hue. It is true he often kept his wallpapers pale in tone, preferring to use the brighter and richer colours for more expensive materials, but he brought into service all colours except those that were 'dingy and muddy', which he was a fault less likely to be curable than that of over-vividness. 'Do not fall into the trap of dingy, bilious yellowy-green, a colour I have a special and personal hatred, because I have somewhat brought it into vogue. I assure you I am not really responsible for it.'

W.S. Gilbert satirised the fashion for Morris's *dirty greens*, and his repudiation of them, in this couplet from *Patience*:

I do not care for dirty greens
By any means.

o-o

By the mid-1870s, some wallpaper manufacturers started to react to public concerns about the use of arsenic. Jeffrey & Co. was amongst those that replaced arsenic greens with less noxious pigments, and was proud to grace its letterheads and advertisements with the phrase 'FREE FROM ARSENIC'. It produced reports from analytical chemists in their literature to verify the slogan, including the following in a handbill:

37, Lamb's Conduit Street, W.C.
July 29th, 1879

To Messrs. (The Manufacturers.)
Gentlemen

I have at your request tested for Arsenic all the colours and metal salts used in your manufactory. I have also examined a number of wall papers (including flocks) taken at random from your stock. I find them all perfectly free from arsenic.

I remain, yours faithfully,
Robert E. Alisom
Late Assistant Chemist to the Royal Arsenal, Woolwich.

Morris & Co. was also pleased to announce that its wallpapers were arsenic-free. Morris's general manager George Wardle wrote a pamphlet describing Morris & Co.'s goods for the Boston, USA, Trade Fair of 1883. 'We may say finally that the colours used in the printing are entirely free from arsenic'.

A critic recounting a display of Jeffrey & Co.'s *Patent Hygenic Wallpapers*, including designs by Morris's acolyte Walter Crane, at the International Health Exhibition in London, 1884, stated 'with our walls covered with such papers we can gratify our artistic taste and at the same time may rest assured that we are not slowly being poisoned'.

Morris & Co.'s and Jeffrey & Co's. competitors William Woolams & Co. were well ahead of the game. Its advertising literature states that it started producing arsenic-free wallpapers in 1859. It is certain that paper-staining manufactures knew about the ill health that their papers created in their staff, let alone their customers.

o-o

How much was William Morris aware of the threats caused by arsenic green wallpaper pigments? A letter from Morris to his dyer Thomas Wardle, George Wardle's brother-in-law, a full decade after his subcontractors Jeffrey & Co. started to advertise 'arsenic free wallpapers', and when even his own promotional material for Morris & Co. was proud to announce that its wallpapers were arsenic-free, reveals his remarkable opinion on the matter.

October 3 [1885]
My Dear Wardle

Thanks for your note about the papers. I cannot imagine it possible that the amount of lead which might be in a paper could give people lead poisoning. Still there should not be lead in them: especially by the way, in the red one: I can understand chromate of lead being in the green ones but surely in small quantities. As to the arsenic scare a greater folly it is hardly possible to imagine: the doctors were bitten as people were bitten by the witch fever. I will see Warner [of Jeffrey & Co.] next week to try to get to the bottom of the matter. My belief about it all is that the doctors find their patients ailing don't know what's the matter with them, and in despair put it down to the wall papers when they probably ought to put it down to the water closet, which I believe to be the source of all illness. And by the by as Nicholson [probably a customer of Morris & Co.] is a tea-totaller he probably imbibes more sewage than other people: though you mustn't tell him I said so.

Yours very truly
William Morris

This letter was followed up a couple of days later by another.

October 6 [1885]
My Dear Wardle

Of course it is proving too much to prove that the Nicholson's were poisoned by wall-papers: for if they were a great many other people would be in the same plight & we should be sure to hear of it. I will get at Warner as soon as I can.

Yrs truly
William Morris

One might have thought that Morris, by this stage of his life an eco-socialist, would have shown some remorse over the probable poisoning of his patrons.

Everyone who read a newspaper regularly knew about poisoning from wallpapers. It was one of the great campaigning issues of the later half of the 19th century. Alfred Swaine Taylor's book *Medical Jurisrudence*, published in many editions throughout the era, stated: 'There appears good reason for believing that a very large amount of sickness and mortality among all classes is attributable to this cause [arsenic wallpaper exposure], and it may probably account for many of the mysterious diseases of the present day which so continually baffle all medical skill.'

Ironically, on a visit to William's childhood home from 1848 onwards, Water House, currently the William Morris Gallery in Walthamstow, I noticed a display board announcing the discovery of some of the wallpaper that was present in the house during William's time. It was discovered under the floorboards in one of the first-floor rooms, a likely place for one of the children's bedrooms. It was a vivid green Jacobean design on a white ground, dated in style to come from the 1840s or 1850s, or perhaps slightly earlier. The bright green looked highly suspicious, and my analysis immediately revealed it to be an arsenic–copper salt, with quite high levels of arsenic. It seems that William might have been a childhood victim, as well as a peddler, of arsenic green wallpaper.

o–o

The mid-19th century saw other arbiters of fashion advocating arsenic wallpapers. Charles's Eastlake's *Hints on Household Taste*, first published in 1868, told the public how to recreate the chic gothic revival style and return to a simpler, more honest design. He suggests that 'very light drab, green (not emerald)' is an ideal colour for wallpaper. In the book, real samples designed for Eastlake, taking inspiration from early Italian wallpapers, are reproduced. They include a rather bilious yellow green with gold embossed stars. My analysis showed that this wallpaper sample contains 0.35 grams of arsenic per page as an arsenic–copper salt. It also had high levels of cadmium, presumably cadmium yellow, to achieve the sickly colour.

By the time the book reached its third edition in 1872 (I have not examined the second edition) the arsenic-stained paper had been removed.

o-o

Between 500 and 700 tons of Scheele's and Emerald green were being produced in England each year according to *The Times* in 1863. Not all of it went into wallpaper. Arsenic greens were also used to dye clothes, paper and cardboard, to colour soap, toys, paint, distemper, artificial and dried flowers, to enhance the look of stuffed animals and green-japanning on tea canisters, and to tint Venetian blinds, curtains and ball gowns. The greens were even used as food colouring in confectionery and puddings. The papers and medical journals were full of arsenic green poisonings by chintzes, cretonnes, tablecloths, lamp shades and silk stockings, all coloured with 'the devil's dust'.

As *The Times* put it:

> The injuries occasioned by the emerald green are not confined to, nor do they indeed chiefly or in any great degree arise from, its poisonous character when swallowed. Mischief does, however, occasionally spring from that cause. The mineral colour is much used in the painting of shops, especially those of bakers, greengrocers, and confectioners. Sometimes the articles dealt in by those persons get stained with the paint, and may be consumed without its being observed. Not only are toys coloured with the arsenical compound, but paper tinted with green is used as an attractive lining for boxes of dried fruit, as a wrapping for chocolate bonbons, and other confectionery, as the material of bags made to contain

> groceries of various sorts, and as a lining for cupboards and drawers, or as covers for tickets, boxes or books. When employed in any of these different ways, the poison is brought within the reach of young children, who not infrequently injure themselves by sucking off the alluring colour. Worse than this, however, confectioners and pastry cooks were, until lately, in the habit of using emerald green as a pigment, and instances are recorded of dangerous or fatal consequences occurring to children from eating fragments of ornamental baskets, cake ornaments, sweetmeats, or apple tarts coloured with poison, and to adults from its ignorant use for the tinting of blanc-mange.

Many candles, nicknamed corpse candles, were made with arsenic. It was added as bleach and a hardener to white ones and as a colour to festive green ones for the Christmas tree. There are reports of people being poisoned by reading books in bed by candlelight.

Fabrics were probably the most dangerous items, as they were worn next to the skin. Green muslin ball gowns, heavy with loosely applied arsenic dye, became so popular they were satirised in W. S. Gilbert's *The Bab Ballad, Only a Dancing Girl* (see Figure 4.4), published in 1869:

> *No airy fairy she,*
> *As she hangs in arsenic green,*
> *From a highly impossible tree, In a highly impossible scene*
> *(Herself not over clean)*

This fashion led an alarmed consulting physician at Guy's Hospital, London, to ask in *The Times* in 1877 about the hazards that green muslin posed to the London social season:

> ...what the atmosphere of a ball room must be where these muslin fabrics are worn, and where the agitation of skirts

Figure 4.4 W.S. Gilbert's drawing to accompany *The Bab Ballad, Only a Dancing Girl.*

consequent on dancing must be constantly discharging arsenical poison. The pallour [sic] and languor so commonly observed in those who pass through the labours of a London season are not to be altogether attributed to ill-ventilated crowded rooms and bad champagne, but are probably in great part owing to the inhalation of arsenical dust

Repeated reports of sickness, discomfort and pustules from wearing such fabrics prompted *The Times* almost a decade earlier to ask 'What manufactured article in these days of high-pressure civilization can possibly be trusted if socks may be dangerous'. The socks in question weren't green: they were magenta. Professor Wanklyn of the London Institution analysed magenta coal tar dyes and found them to be pure arsenate salt of an aniline dye.

o-o

John Glaister, the Professor of Forensic Medicine at Glasgow University, could not quite understand a Scottish aversion to green-coloured confectionery. In his 1954 book *The Power of Poison* he has this observation:

> Very recently, it was reported in the press that a speaker at a meeting of confectioners in London had said, 'Fewer green-tinted sweets are sold in Scotland than in any other country'. Another speaker, who had owned a Glasgow confectionery business for thirty-seven years, had told a reporter that, 'Even green decorations on an iced cake can reduce the possibilities of sale in many parts of Scotland. I have often had to eat one myself because my customers preferred to go without rather than touch the arsenical-looking things'.... It is astonishing why the idea of green colour in confections should suggest the stupid impression that arsenic is present.

Professor Glaister did not do his homework. There was a very good reason why the Scots were terrified of green confectionery. Arsenic greens were regularly used in Scotland in the 19th century as a food colouring. Greenock, in Glasgow, was extremely proud of its name. It had a green fetish, 'of terrible fascination, seen on every wall and counter in the town', said one commentator. Arsenic greens were to be found on flowers, feathers, bonnet ribbons, tapers and candles. Most disturbingly, a Christmas cake confiscated by the authorities from window display, embedded with a green card coated with sugar reading 'For the bairnies' contained seven times the fatal adult dose.

The Irish penchant for green also had grisly consequences. After the Irish Crimean banquet in London, where Scheele's green was used to provide a patriotic backdrop, several of the guests took the confectionery decorations home to their families, resulting in a child's death.

The practice of using arsenic greens in confectionery seems to have been extraordinarily widespread, as evinced by this report from the New York satirical magazine *Puck* from 1885 (see Figure 4.5):

> The adulteration of food has long been discussed in the papers as a piece of deviltry that should be stopped by the imprisonment of the perpetrators. But this crime against the public stomach has lately been eclipsed by fiends who are said to have used poison in the construction of candy that glads the eye of childhood, and gild the golden vistas of love's young dream. At any rate, vast quantities of the highly-coloured stuff have been dumped into the rivers. This seems wrong, because the candy might poison the fish, and the fish would, in all probability, poison many of the faithful during the Lenten season.

Figure 4.5 Confectionery cartoon from *Puck* magazine, 1885.

> We think it would be better to stock our forts with it, to mix with grape and canister. The candy-poison scare will, no doubt, be utilized profitably by all young men who are engaged to be married, while any modern Lucretia Borgia may give her victim taffy with success.

o-o

The British government was slow to investigate the arsenic complaints, and slow to act when it found a problem. Following the 1851 Arsenic Act, which restricted the sale of white arsenic, no legislation was put into place to control the arsenic trades or to protect the public from poisonings until 1895. In 1883 Lord Granville, Gladstone's Foreign Secretary, published a survey of overseas laws covering the manufacture, use and sale of arsenical pigments. It concluded that Britain was not backward in lacking legislation on arsenic greens, as there was none in the USA, Belgium, Greece, Italy, the Netherlands, Portugal or Romania.

The most stringent regulations were in Sweden and the German-speaking states. The Germans had banned the use of 'poisonous colours for paper-hangings and wearing apparel' in 1879, singling out arsenic pigments in particular. Arsenic in paper hangings had been troubling the Germans for some time; it appears that this ban had its origins in an Order published in 1854 in Prussia stating: 'Paper-hangings printed with colours containing arsenic are only to be allowed in trade with foreign countries'. Sweden banned the import and manufacture of arsenic-coloured goods from 1876, it concluded that such objects were particularly dangerous to Swedes: 'Living, as they do in winter, in almost hermetically closed apartments'.

o-o

In some ways the medical press confused matters considerably. While campaigning against the wallpaper problem, they were under the spell of economics. As *The Lancet* put it in an article on arsenic green pigments:

> While it is the duty of the sanitarian to guard with jealous eye the public health, he should be most careful that in doing so he does not unnecessarily interfere with trade and manufactures.

What's more, Victorian Society was fascinated by the health tonic Fowler's solution and stories of the remarkable arsenic eaters of Styria in Austria (of which both, more later). Alongside articles decrying arsenic pigments, vapours and dust, *The Lancet* hailed arsenical cigarettes for stopping nervous spasms.

Such promising claims were also published in popular periodicals, including a series on the use of arsenical pigments in wallpapers by Mathieu Williams in *The Gentleman's Magazine* from 1881 onwards:

> I have long held very heretical opinions on the subject of poisoning by arsenical wall-papers, even going so far to believe that, if they have any effect at all, it is beneficial, on account of the powerful disinfectant properties of very small quantities of arsenic and of arsenical vapours.

Williams went on to argue that Birmingham had become free of cholera due to arsenic emitted from copper smelting, and that if he lived in New Orleans or other fever hotspots he would envelop himself in 'arsenical vapours', by covering his room in wallpapers and upholstery and carrying arsenic-dosed handkerchiefs. He even suggested that fever hospitals should be supplied with arsenic wallpapers. Williams himself sickened during a stay in a hotel room lined with green flock wallpaper. Even this experience did not diminish the vigour with which he promoted his bizarre philosophy.

o–o

Eventually commonsense had to intervene. The Factories and Workshop Act of 1895:

> ...obliged every medical practitioner attending on or called to visit a patient suffering from poisoning by lead, phosphorus or arsenic, or from anthrax, contracted in a factory or workshop, to notify it to the Chief Inspector of Factories.

By this time customers had already started to spurn arsenic greens. A *Special Analytical Commission on Arsenical Wallpaper* in 1892 found:

> Prejudice against a design in green rose high for reasons already cited and in the minds of many green is still the colour to avoid and the one most to be suspected. But on this very account more than any other, with the idea of substituting for it a material equally effective and elegant, but free from any substance which could be regarded as injurious.

In 1896 a Home Office interim report into *Certain Miscellaneous Dangerous Trades* concluded:

> The industry by which wall papers are prepared is one which has undergone very considerable change in the last 25 years. Attention was first called to it in 1839, and since then, from time to time, there have been many serious and justifiable complaints, not only from the operatives engaged in the trade, but from the public utilising the manufactured article.

The report noted that the industry had by then largely reformed itself and had substituted vegetable dyes for arsenic ones.

o–o

From a time when systematic medical surveys were far from the norm, one piece of evidence shows the scale of Victorian domestic arsenic poisoning.

A Dr Putnam, interviewed for the *Chicago Journal* in 1891, set about finding how prevalent arsenic exposure was in the home. He analysed urine samples from 150 patients (the kidneys rapidly excrete ingested arsenic). The patients came from a wide range of social backgrounds, and some of them presented obscure illnesses, while others were selected because they had no symptoms that could be related to poisoning. Thirty per cent had arsenic in their urine.

Dr Putnam concluded from these observations that his community was 'exposed to arsenical contamination on a very large scale' and that this arsenic exposure was from 'domestic conditions'. Although his trial was biased by selecting sick patients, the fact that one in three people had high levels of this most notorious of poisons in their bodies was startling. Dr Putnam's survey was not repeated in other conurbations, but the mass-produced goods that caused this wide-scale arsenic poisoning in Chicago were to be found in every region of the world subjected to Victorian commerce.

Chapter 5
HEALING ARSENIC

Poisons in small doses are the best medicines, and the best medicines in too large doses are poisonous.

William Withering, *Botanical Arrangement of all the vegetables naturally growing in Great Britain*, 1776.

Arsenic compounds have been used in medicine from ancient times till today. During the 19th century, arsenic became fashionable as an ingredient of cure-all tonics. This infatuation led to the mass poisoning of the Victorian world through medication, adding to the mass poisoning from interior décor.

In the 18th century doctors prescribed arsenic for skin diseases, neuralgia, fever, malaria, syphilis, lumbago and epilepsy. It gradually gained a reputation as a panacea. Of course, arsenic does destroy a wide range of pathogens, and so probably helped get rid of some infectious diseases; but a cure that kills the patient is not much of a remedy. In an age in which clinical trials were at best rudimentary, the toxic side effects of a drug could be safely ignored in the pursuit of profit.

A history of arsenic as a medicine was published in *The Lancet* in 1837 by George G. Sigmond MD. He cites a paper read at the College of Physicians on 10 January 1785:

> ...that a medicine compounded of arsenic and opium, the dose of which is a very few drops in water, was taken by some of the inferior ranks of people, and sometimes successfully, but, now and then, violent vomitings, colics and dysentery, were the effects of it.

Many minor quacks peddled poisonous potions. Up to sixty arsenic-based medicines claimed therapeutic properties, including Aiken's Tonic Pills, Andrew's Tonic, Arsenauro, Gross's Neuralgia Pills, Chloro-Phosphide of Arsenic, Sulphur Compound Lozenges, De Valgin's mineral solution, Donovan's solution, Asiatic Drops and Asiatic Pills. But one man, Dr Thomas Fowler, born in Salisbury, England, dominates the story of arsenic in 19th medicine. He published his findings on arsenic in an advertising pamphlet *Medical Reports of the Effects of Arsenic in the Cure of Agues, Remitting Fevers and Periodic Headaches*, in 1786.

Working in Stafford Infirmary with the apothecary Mr Hughes, Fowler analysed a patent medicine: Thomas Wilson's *Tasteless*

Ague and Fever Drops. The two men discovered that the active ingredient was arsenic. Fowler then developed his own *Liquor Arsenicalis*, a 1% solution of potassium arsenite, coloured with a tincture of lavender. A few drops of this solution were taken diluted in water. Fowler's solution, as came to be called, entered the London Pharmacopoeia in 1809 and remained in Western formularies until the 1970s. It contained a massive amount of arsenic: 5,000,000 ppb. In contrast, the maximum level found in West Bengal and Bangladesh is 4,000 ppb. Fowler's treatment involved 24 doses at a rate of three per day, with a dose being 12 drops. The total course of 280 drops was equivalent to just over 0.06 of a gram of arsenic. The maximum recommended single dose was 0.5 ml of the solution – 3 mg of arsenic. A fatal dose of potassium arsenite is between 36 and 108 mg. A patient taking three maximum doses per day would have consumed 9 milligrams of arsenic.

Thomas Fowler was a medical student at Edinburgh University, famed for its progress in 18th and 19th century medicine. Here he was able to secure the patronage of powerful allies such as Andrew Duncan, Physician to His Royal Highness the Prince of Wales for Scotland, to whom Fowler dedicated his arsenic study. For marketing reasons he ensured that 'arsenic' was not used in the name of his medicine: 'the idea of a poison seems to be so strongly connected with that of arsenic, it will be very difficult to separate them in the mind'. He instead called his concoction 'Mineral Solution'. Later, the solution became associated with his name.

What characterised Fowler's approach to medical treatments, including arsenic, was the large number of case studies that he presented in his advertising pamphlets. These careful observations were an early example of systematic pharmacological investigation. Fowler noted the good and the bad in his experiments, writing that one third of his patients experienced 'operative effects', including nausea, vomiting, diarrhoea and abdominal pain. Fowler believed that such side effects were beneficial.

In total he tested his solution on 247 cases. He 'cured' a total of 171 patients, though 27 had relapses. His arsenic treatise reports a total of 86 individual case histories. Here is the last:

> Sarah Eldershaw, of Stafford, aged 23, out-patient, was affected with a quotidian headache, of six days continuance, and had the paroxysms rendered milder by taking fourteen drops of the solution, three times a day, for three days. It operated as a slight emetic, attended with nausea and griping. She was ordered to repeat the medicine for four days; but made no report from a neglect of attendance.

Thomas Fowler set out, right at the start of his arsenic tract, the resistance that his formulation might face:

> Arsenic is a mineral which has long been reputed one of the most violent poisons hitherto known; and accordingly has been reprobated in the strongest terms by almost every medical writer, that has ever deigned to notice it; ... and yet there is good reason to believe it bids fair to hold a place, among the best and most valuable medicines; and to rank with the Peruvian Bark [quinine] in the cure of agues, remitting fevers, and periodic headaches.

Peruvian Bark was expensive; arsenic was cheap and plentiful. Arsenic rapidly became the 'cure' for malaria. Later, it was used to treat eczema and psoriasis, asthma, chorea, rheumatic fever, pain, anaemia, leukaemia and Hodgkin's disease.

Praise for Fowler's solution was fulsome. In 1858, James Begbie, Vice-President of the Royal College of Physicians of Edinburgh, published a paper in the *Edinburgh Medical Journal* concerning *Liquor Arsenicalis*:

> Arsenic, I believe, is now much more extensively employed in the treatment of disease... and much of the dread and

> apprehension of its poisonous effects, which deterred many from its use altogether, or suffered its exhibition to be so limited as to cause it to fall short of the production of physical operation, have happily been overcome; and the profession now generally regard it as one of the most useful and available of its therapeutic agents, powerful in many intractable affections, and exercising a commanding influence over ailments hitherto considered incurable.

Begbie knew the side effects of taking Fowler's solution, but he urged his readers to persist nonetheless:

> Arsenic, when given continuously in moderate doses – say five drops of the liquor arsenicalis, diluted largely in water, twice or thrice a-day – will, sooner or later, generally within eight to ten days, produce increase of heat and dryness of skin, together with acceleration of pulse, followed by sense of heat and itchiness of the eyelids, to which succeed swelling and tenderness; the conjunctiva becomes inflamed, the eye sensitive to light, and the orbit surrounded by a dark colouration. The tongue at this time will be found finely coated with a white silvery film.... The throat becomes dry and sore, the gums swollen and tender; and if the remedy is still further persisted in, salvation ensues. Nausea, vomiting, diarrhoea, nervous depression, faintness and tremor are added to the catalogue of ills: but the judicious practitioner will suspend the drug long before these symptoms show themselves.

As Begbie was Queen Victoria's physician in Scotland, she may have been an imbiber of arsenic. Victoria's political antithesis, Karl Marx, definitely did partake. He went to Margate, England, to find relief from carbuncles and liver problems, and was prescribed a solution of arsenic three times a day. Later, following another attack of carbuncles, he eschewed the treatment, remarking that arsenic 'dulls my mind too much and I needed to keep my wits about me'. One of Marx's heroes, Charles Darwin, also took arsenic to treat eczema of his lip and a skin problem on his hand.

One I.L. Crawcour, typifying the 19th century physician's reliance on arsenic, wrote in the *Journal of the Louisiana State Medical Society* in 1883: 'If a law were passed compelling physicians to confine themselves to two remedies only in their entire practice, arsenic would be my choice for one, opium for the other'. Fowler's solution became the therapeutic mule for 19th century medicine.

o-o

The lasting effects of ingesting Fowler's solution for long periods were terrible. Caution regarding arsenic as a medicine was sounded as early as 1856, when an anonymous letter appeared in *The Lancet*: 'Sir – As of late several cases have come under my observation, in which the long-continued use of arsenic in minute doses has been productive of very serious evils, I am desirous of knowing whether any of my professional brethren have met with such cases'.

Cases of skin cancer and keratosis associated with arsenic medication were published in 1888 by Jonathan Hutchinson in a paper to the *Transactions of the Pathological Society of London*: 'In the following statement I have two separate propositions to maintain.... The first is that by prolonged internal use of arsenic the nutrition of the skin may be seriously affected, and that, amongst other changes, warty or corn-like indurations may be produced. The second goes much further and asserts that if the drug be continued these "arsenic corns" may assume a tendency to grow downwards and pass into epithelial cancer'.

Hutchinson could be describing the symptoms seen today in Bengal. He observed patients who had all taken Fowler's solution over years, at high doses, for psoriasis. He published pictures of the keratosis and skin cancer (Figure 5.1).

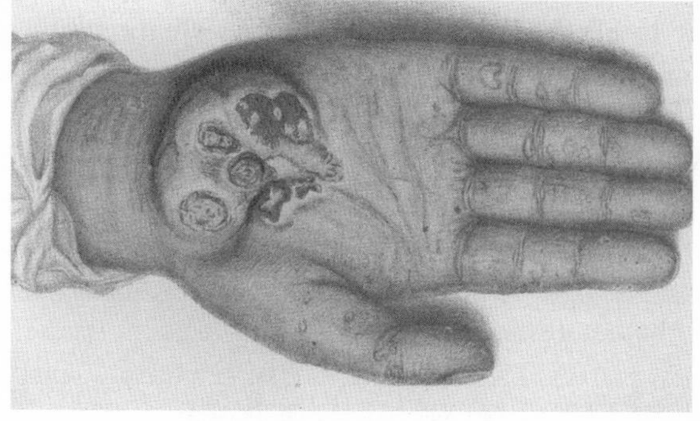

Figure 5.1 First published drawing of arsenic-induced cancer, by Jonathan Hutchinson in 1888.

The patient with skin cancer had his arm amputated one month after attending Hutchinson's clinic and was dead of internal cancers 16 months later.

William Osler's medical text *The Principles and Practice of Medicine*, first published in 1894, records that patients taking 15–20 drops of Fowler's solution three times a day over years developed 'cutaneous pigmentation and keratosis'. By the 5th edition of 1903 the Manchester arsenic beer debacle had occurred, and Osler noted that in 'nearly all cases' patients showed signs of 'lesions of the skin'.

As late as 1902, Professor Stockman of Glasgow University wrote in the *British Medical Journal* that the ill effects of Fowler's solution 'can hardly be described as more than a temporary inconvenience'. However, the 'medicine' was finally withdrawn from the US market in the 1950s. One case history published in the *Journal of the American Medical Association* in 1952 has a picture of a Fowler's solution victim with his chest covered in the black rain common in tubewell victims.

The long-term effects of arsenic exposure were too subtle for most 19th century physicians to fathom, but some treatments resulted in rapid death. Arsenic poultices applied to wounds often proved fatal. Writing about these treatments, the author of a 19th century pharmacopoeia, John Ayrton Paris, warns that 'repeated experiments have proved that arsenic kills'. This did not stop anyone marketing arsenic to treat external cancers. Paris lists some of these 'cures':

> PLUNKETT'S OINTMENT, consists of arsenious acid, sulphur and the powdered flowers of Rannunculous Flammula, and Cotula Faetida, levigated and made into a paste with the white of an egg, and applied on a piece of pig's bladder to the surface of the cancer.
>
> PATER ARSENICALE. This favourite remedy of the French surgeon consists of 70 parts of cinnabar, 22 of sangius draconis and 8 of arsenious acid, made into a paste with saliva, at the time of applying it.
>
> DAVIDSON'S REMEDY FOR CANCER, arsenious acid and powdered hemlock.

o–o

Paris then goes on to give a cosmetic application of arsenic:

> DELCROIX'S POUDRE SUBTIL, 'for removing depilatory hairs in less than ten minutes'! This fashionable depilatory appears upon examination to consist of Quicklime and Sulphuret of Arsenic, with some vegetable powder. It is however, so unequally mixed, that in substituting it to analysis, no two portions afforded the same results. It can scarcely be necessary to state, that such a composition is incapable of fulfilling the intention to which it so confidently intended.

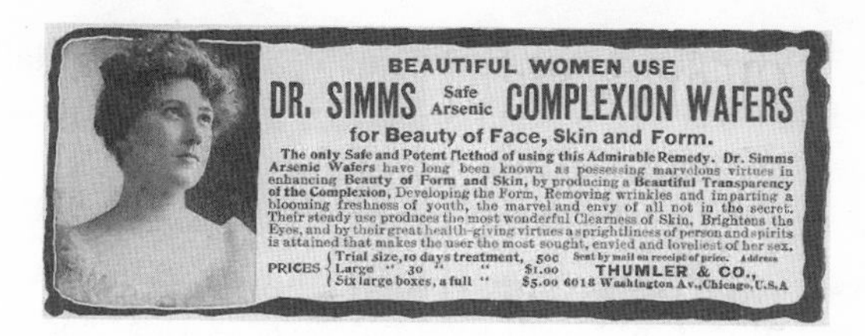

Figure 5.2 An advert for an arsenic-based beauty product.

Fowler's solution too was marketed, not just as a medicine but as a tonic to improve the skin and increase vitality. It certainly brought about the 'milk and roses' complexion sought by refined Victorian ladies. The roses could be obtained by ingesting arsenic, which dilated the blood vessels of the skin. The milk could be brought about by using arsenic trioxide as a wash. It bleached the skin, as this 1818 observation of the arsenic miners of Saxony illustrates: 'The white [arsenic trioxide] is particularly dangerous to handle, being of the most poisonous nature. I observed running water throughout the works, and the frequent use of it made by the workmen was evident from the lady-like whiteness of their hands'.

Arsenic was also used to delouse hair, as wig powder, and in soaps. Women's magazines carried adverts for it (Figure 5.2).

o–o

In the 1850s arsenic's reputation as a cure-all received a massive boost. Victorian society became fascinated with tales emerging from the Austrian Alps about the arsenic eaters of Styria. The story broke in the *Chambers Edinburgh Journal* (a popular magazine reporting on culture and science) in an article entitled 'The Poison Eaters':

> A very interesting trial for murder took place lately in Austria. The prisoner Anna Alexander, was acquitted by the jury who, in the various questions put to the witness in order to

> discover whether the murdered man, Lieutenant Mathew Wurzel, was a poison-eater or not, educed some very curious evidence relating to this class of persons.

The poison in question was arsenic trioxide. A series of articles in *Chambers* and its Edinburgh rival *Blackwoods Magazine* caused a considerable stir, leading to reports appearing in the popular and scientific press.

The tales of arsenic eating were published the 1855 book *The Chemistry of Common Life* by J.F.W. Johnston, then Professor of Chemistry at Durham University. Like *Chambers* and *Blackwoods*, Johnston reported that Styrian peasants regularly consumed large quantities of arsenic to give plumpness to their figures and cleanliness, softness, beauty and freshness to their complexion. They felt it aided digestion, acted as an aphrodisiac and improved their stamina. Austrian mountaineers apparently used it to boost their climbing capabilities. It was thought that stablehands adopted the practice after seeing the improvement in horses fed arsenic.

Styria is in the hilly south of Austria near Hungary. Inhabitants harvested white arsenic trioxide from the soot that gathered in the chimneys used to smelt arsenic-containing ores. The white arsenic was known as *hidri*, from *hütten-rauch* or smelt-house smoke. Arsenic eating was done in secret, as its magical whiff was unpopular with the church. This secretiveness made the phenomenon difficult to verify, and its revelation to the world met with scepticism.

Styrians ate their arsenic with bread and bacon. They apparently built up immunity slowly, taking frequent small amounts and raising the dose over years. *Hidri* had a stable and reliable concentration of arsenic trioxide that the gut would have absorbed slowly. They started taking 10 mg per day, which they increased every couple of days to get up to 300–400 mg. Normally the fatal dose of arsenic trioxide is 70–180 mg. Poisonings were rare and arsenic eaters could take high doses over a 30–40 year period.

Accidents did happen though, as Johnston's book recalled:

> A healthy, but pale and thin milkmaid, residing in the parish of H---, had a lover whom she wished to attach to her by a more agreeable exterior; she, therefore, had recourse to the well-known beautifier, and took arsenic several times per week. The desired effect was not long in showing itself; for in a few months she became stout, rosy-cheeked, and all that her lover could desire. In order, however, to increase the effect, she incautiously increased the dose of arsenic, and fell victim to her vanity. She died poisoned, a very painful death.

An arsenic eater who stopped developed symptoms of slight arsenic poisoning – anxiety, upset digestion, loss of appetite, increased salivation, burning in the stomach, spasms in the throat, pains in the bowels, constipation and difficulty breathing. The only speedy mode of relief was to eat arsenic again.

o–o

More reputable uses of arsenic as a medicine developed alongside the cure-alls and cosmetics. In 1865 the German physician H. Lissauer used arsenic as the first chemotherapy agent to treat leukaemia, a disease first described in 1845 by John Hughes Bennet of Edinburgh. It produced a transient recovery. In 1931 Claude Forkner and McNair Scott of Boston City hospital rediscovered Fowler's solution, in the treatment of chronic myeloid leukaemia. Arsenic compounds and irradiation remained the therapy of choice for leukaemia patients until busulfan was introduced in 1953. Today, much leukaemia research is focused on arsenic-based drugs, including arsenic trioxide. A drip of 10 milligrams of arsenic trioxide for 45 days induces the remission of acute promyelocytic leukaemia, apparently with no side effects.

Figure 5.3 Paul Ehrlich.

Organic arsenic compounds, where the element is incorporated into a carbon molecule, were the first antibiotics produced by targeted synthesis in the laboratory, heralding the birth of modern pharmacology at the turn of the 20th century. They proved to be the only drugs effective against the debilitating and deadly African sleeping sickness, trypanosomiasis, transmitted by the tsetse fly and widespread in sub-Saharan Africa.

These organo-arsenic molecules were developed by Paul Ehrlich (Figure 5.3). His first success was the compound Salvarsan or 606, Ehrlich's famous 'magic bullet'.

Ehrlich was born into a prosperous Jewish family in Strehlen, Upper Silesia, on 14 March 1854. He had an early interest in chemistry and at university became involved in the burgeoning field of microscopy, helping to develop a number of stains for biological tissues, including a diagnostic one for the bacterium that causes tuberculosis. Ehrlich obtained his Doctorate in Medicine in 1878 with the very first analytical study of staining methods, investigating the properties of aniline dyes. It was these early studies into the binding of chemicals to biological tissues that inspired his life's work – 'the production and introduction into medicine, of chemical compounds which act as specifics for human diseases due to infections'. He hit on the idea that one could inactivate a physiological process or kill a parasite, say, by finding or designing chemicals to react with receptors in the cells.

By the early 1900s physicians were using arsenic compounds such as sodium cacodylate to treat syphilis, pellagra, malaria and sleeping sickness. In 1906 Harold Wolferstan Thomas and Anton Breinl working in Liverpool's School of Tropical Medicine found that the arsenic compound atoxyl cured trypanomiasis. Unfortunately, atoxyl also destroyed the optic nerve and had to be abandoned for fear of patients becoming blind.

Thomas and Breinl inspired Ehrlich to synthesise compounds with similar curative properties to atoxyl but with less toxic side effects. He struck lucky with his 606th preparation, dioxy-diamino-arseno-benzene, commercially named Salvarsan – meaning 'healing arsenic'. It was first used successfully in humans in 1911. The German chemical giant Hoechst gained approval to market Salvarsan and distributed 65,000 free samples throughout the world in return for information on patients' response.

Salvarsan also proved to be effective against syphilis. Before preparation 606, syphilis treatment was very unpleasant, inspiring the refrain: 'A night with Venus, a lifetime with mercury'. Mercury gave only a respite from syphilis, not a cure, and had very unpleasant side effects. Salvarsan was not much better, but at least it controlled the disease. Judith Thurman's biography of Karen Blixen, the author of *Out of Africa*, details her long fight against syphilis and the side effects of arsenic drugs.

Later, Ehrlich came up with less toxic derivatives of Salvarsan: Neosalvarsan (or 909), Oxophenarsine hydrochloride (or Mapharsen) and Silver arsphenamine. In total, more than 32,000 arsenic-containing compounds, the vast majority of which did not make the grade, have been synthesised in the search for drugs, more than for any other trace element.

Organic arsenic compounds remained the mainstay of syphilis treatment for nearly 40 years until the invention of penicillin. They are still used to treat sleeping sickness, despite an estimated 1,000 deaths each year from side effects. There are no alternatives.

o-o

Remarkably, the Scottish African adventurer Dr David Livingstone was the first to use arsenic compounds in the treatment of trypanosomiasis, some fifty years before Ehrlich. His experiment on an infected mare was published in the *British Medical Journal* in 1858, reporting events a decade earlier when he was based in Kolobeng, South Africa, during 1847 and 1948:

> A mare belonging to Mr. Gordon Cumming was brought to Kolobeng, after prolonged exposure to the bite of the insect; as it was unable to proceed on the journey southwards, its owner left it to die. I gave it two grains of arsenic in a little barley for about a week, when an eruption resembling small-pox appeared. This induced me to discontinue the medicine; and when the eruption disappeared the animal's coat became so smooth and glossy that I imagined I had cured the complaint; for, after the bite is inflicted, the coat stares as if the animal were cold.
>
> ... About two months after this apparent cure, the coat began to stare again, but this time had a remarkable dryness and harshness. I tried the arsenic again; but the mare became like a skeleton and refused to touch the barely. When I tried to coax her, she turned her mild eye so imploring and so evidently meaning, 'My dear fellow, I would rather die of the disease than of the doctor', that I could not force her. I got her lifted every morning to feed, and saw her

at last perish through sheer exhaustion; and this nearly *six months* after the bite was inflicted.

o–o

There are surprisingly few detailed epidemiological studies of the long-term health consequences of taking inorganic arsenic. The only one I could find was a survey of Swiss patients taking Fowler's solution for between 6 and 26 years. Dr Fierz, working in a Zurich medical clinic, came across a junior doctor whose general practitioner father had been treating skin disease with Fowler's solution. Fierz sought permission to write to 1,450 of his patients, asking them to attend a clinic for inspection; 262 turned up between March 1963 and March 1964. Fierz examined only those under the age of 65, to rule out age-related cancers. Most patients said that the Fowler's solution helped their skin. Unfortunately, they were swapping diseases like eczema for pre-cancerous and cancerous growths.

Of the patients, 106 (42%) had keratosis on their palms. In the main, the patients had failed to notice these 1–3 millimetre wide blemishes, or had assumed that they were corns. Only five patients had hyperpigmentation. But 8% had skin cancer. This is usually found on skin exposed to lots of sunlight – the hands and face, say. Fierz's patients had lesions on their trunks, an area normally shielded from cancer-causing ultraviolet light.

Fierz calculated that the keratosis had a latency of 6 years, and the skin cancer had a latency of 14 years. The most severely affected patients were those with psoriasis. They had taken higher doses of arsenic, over longer periods, because of the persistent nature of their disease.

This Swiss study shows that arsenic-related illness is not just a developing world problem. Switzerland is a rich country, even by western standards. It appears that good nutrition, which is often cited as one of the factors mitigating the poison's effects, can at best only delay them.

Fierz's study also highlighted something more worrying: secondary cancers developed in the lymph glands in some of the patients, giving rise to internal cancers.

Chapter 6
TO FRUSTRATE THE AIM OF JUSTICE

Given the patient's history, that is, the old man's suspicions, it would have been imprudent to consume the sugar blindly, even just taste it. I dissolved a little in distilled water: the solution was turbid – there was something wrong with it. I weighed a gram of sugar in the platinum crucible (the apple of our eyes) to incinerate it on the flame: there rose in the lab's polluted air the domestic and childish smell of burnt sugar, but immediately afterward the flame turned livid and there was a much different smell, metallic, garlicky, inorganic, indeed contra-organic: a chemist without a nose is in for trouble. At this point it is hard to make a mistake: filter the solution, acidify it, take the Kipp, let hydrogen sulphide bubble through. And here is the yellow precipitate of sulphide, it is arsenious anhydride – in short, arsenic, the Masculinium, the arsenic of Mithridates and Madame Bovary.

Primo Levi, *The Periodic Table*, 1975

Poisoners fascinate us. Mountains of true crime books pick over the bones of the most notorious cases. And fictional poisonings are the crime writer's staple. It is no surprise, then, that arsenic is central to many true and imaginary murders.

Indeed, white arsenic – arsenic trioxide – has been called the perfect poison. It is nearly tasteless or slightly sweet, colourless, odourless and slightly soluble. Half a teaspoon in a cup of water, wine or tea can kill. It made crime so simple that even an old lady like Aunt Martha in the Broadway hit and film *Arsenic and Old Lace* (Figure 6.1) could indulge: 'Well, dear, for a gallon of elderberry wine I take one teaspoon of arsenic...'.

o-o

Until the development of forensic tests for arsenic, murderers could poison large numbers of people without detection. As we have seen, the symptoms of arsenic poisoning resemble cholera, common in the West until the second half of the 19th century.

Sir Thomas Overbury, for instance, died in 1613, apparently from natural causes, while imprisoned in the tower of London for court intrigues. A deathbed confession two years later by an apothecary's assistant led to an inquiry and trial. The apothecary, James Franklin, eventually admitted to supplying white arsenic, which he said was put in the prisoner's food by the gaoler, on the instructions of the Countess of Essex. Franklin and four others were executed. The Earl and Countess of Essex were convicted but subsequently pardoned, leading to suspicions that King James I was involved in the plot.

While the English dabbled in the black art of arsenic poisoning, the French and Italians revelled in it. Catherine Deshayes, also known as La Voisin, meaning The Neighbour, had a commercial poisoning business during the reign of Louis XIV (1638–1715). From her Paris headquarters, beside the church Notre-Dame-de Bonne-Nouvelle, La Voisin made a powder of arsenic trioxide and toxic vegetable extracts, such as aconite, belladonna and

Figure 6.1 Playbill from the original Broadway production of *Arsenic and Old Lace*.

opium, that became known as *poudre de succession*, as it reputedly allowed many heirs to take their titles at an early age. She was accused of poisoning 2000 infants and numerous adults. La Voisin and her accomplices were eventually arrested and executed on 22 February 1680. The spree shocked the French court, as La Voisin was apparently consulted, because of her reputation for witchcraft, by Madame de Montespan, Louis XIV's mistress. De Montespan wanted a love potion to entice the King back to her bosom. The potion comprised dried moles and bat's blood.

Another notorious French poisoner was the Marquise de Brinvilliers. She too terrorised Paris during the reign of the Sun King. The married Marquise took a lover, Gaudin de Sainte-Croix; her father disapproved and had Sainte-Croix imprisoned. On release Sainte-Croix and his mistress set about retribution. Legend has it that they refined the art of poisoning by distributing arsenic-laced jam pastries as gifts to the sick of Paris's hospitals, with no one suspecting the already ailing patients to be dying of anything untoward.

When the pair found a dosage that singly was not fatal, but which resulted in a long and terminal illness if repeated, they set out to bump off the Marquise's father by degrees. Fed poisoned food, the old man got ever more sick, all the time so well cared for by his dutiful daughter that he changed his will to reward her devotion. Once her father died, the Marquise went about killing her two brothers to inherit their wealth. Autopsies of her father and one of the brothers found no evidence of foul play. A post mortem of the second brother revealed the characteristic stomach ulceration of arsenic poisoning, but the Marquise did not come under suspicion. Next on her list was her husband, but she took pity on him as his condition worsened, and ceased administering the poison.

Justice caught up with the Marquise when Sainte-Croix suddenly died in 1672, leaving incriminating letters. In 1676 she was arrested, beheaded and her body burnt.

An interesting aside to this story is that at one point in her letters to Sainte-Croix the Marquise threatens to use the 'recipe of Glaser' to commit suicide. Glaser was a Swiss alchemist with a shop in Paris. He was also Professor at the Jardin du Roi, and Apothecary to the King and the Duke of Orleans. Glaser wrote a treatise on chemistry that includes a recipe for 'a corrosive liquor of arsenic'. He is thought to have supplied the Marquise, and spent some time in the Bastille for his dodgy dealings.

The Italians had their counterpart in Signora Tophana of Sicily, who began her career as a poisoner in Palermo in 1650. In 1659 she moved to Naples and built an underground distribution network for the lethal potion she developed. For over 50 years Tophana purveyed her blend of foxglove extract and an arsenic trioxide under the names *Aqua Toffania*, *Aqua della Toffana*, *Acquetta di Napoli* or simply *Acquetta*. Four to six drops were sufficient to kill a man, and the dose could be varied as to operate within any given period. To escape detection, Tophana packaged her wares in small glass phials, inscribed, 'Manna of Saint Nicolas Bari' ornamented with an image of the local saint. Oil or water alleged to emanate from Nicolas Bari's grave was considered a miracle cure. Put to the rack in 1709, Signora Tophana confessed to killing more than 600 people, for which she was condemned to death by strangulation.

⊶

Unveiling the wickedness of 17th and early 18th century poisoners needed confessions or incriminating evidence, so many an arsenic crime probably went unnoticed. To bring these dastardly deeds to light, forensic chemistry was necessary. Since arsenic can readily be converted into a gas, it was easy to separate and concentrate from liquids or tissues. Minute amounts could be detected with simple equipment.

One of the first triumphs for forensic science came in 1752. Mary Blandy, a middle-class educated lady, the daughter of a lawyer and Town Clerk from Henley on Thames in Oxfordshire, mixed white arsenic with her father's food on the instructions of her lover. At Mary's trial, Dr Addington, the medic who attended her father, told the jury:

> ...after drinking more gruel on Tuesday night August the 6th, he had felt the grittiness in his Mouth again, and that the burning and prickling in his tongue, throat stomach and bowels, had returned with double violence and had been aggravated by a prodigious swelling of the belly, and exquisite pains and pricklings in every external as well as internal

> part of his body, which pricklings he compared to an infinite number of needles darting into him all at once.

Later in the trial Addington went on:

> When I got downstairs one of the maids put a paper in my hands which she said Miss Blandy had thrown into the fire, several holes were burnt in the paper but not a letter of superscription was effaced. The Superscription was, 'The Powder to clean the Pebbles with'.... I opened the paper very carefully, and found in it a whitish powder, like white arsenic in taste, but slightly discoloured by a little burnt paper mixed with it.

Dr Addington performed simple tests on the colour and texture of the white powder, and guessed it to be arsenic trioxide. Mary Blandy's loving father, when the truth of his poisoning was revealed to him shortly before his death, didn't want her condemned to death, but the public bayed for blood. She was convicted and hanged on Addington's evidence.

o-o

Techniques for detecting arsenic improved at the start of the 19th century. In 1809 the German Chemist Edgar Hugo Emil Reinsch discovered the following way to spot traces of arsenic in contaminated liquids such as poisoned drinks:

> A slip of copper-foil boiled in poisoned liquid, previously acidulated with hydrochloric acid, withdraws the arsenic and becomes covered with a white alloy. By heating the metal in a glass tube the arsenic is expelled.

This assay was widely used, despite one limitation. It required very pure copper, as lower grade metal often contains arsenic.

In the early 1830s Professor Jöns Jacob Berzelius, a Swedish chemist, developed another test. The stomach and intestines of the postulated murder victim were removed and dissolved with potassium hydroxide. Hydrochloric acid was then added and the resulting solution was filtered and concentrated by evaporation. Hydrogen sulphide was then passed through the solution. If the deceased had ingested arsenic the yellow arsenic–sulphur compound orpiment would precipitate at the bottom of a thin glass tube.

Also developed in the 19th century was the Marsh test. This remained the standard assay for arsenic until the 1970s. The stimulus for developing the test, first described in the *Edinburgh Philosophical Journal* in 1838, was the acquittal of John Bodle, who was charged with the murder of his grandfather George Bodle in 1832. Bodle Junior later admitted to the crime and sold his story to the newspapers, greatly irking chemist James Marsh, who at the trial was asked to prove, if he could, the presence of arsenic in some suspect coffee and in the organs of the dead man. Marsh had used the Berzelius test. The telltale yellow precipitate did indeed form when he bubbled hydrogen sulphide gas into the solutions in question, but the colour dissipated with time and the jury was not impressed.

Marsh worked under Michael Faraday at the Royal Arsenal in Woolwich, the Crown's laboratories for developing and storing munitions. His new test involved reacting any inorganic arsenic in solution with hydrogen to give the garlicky gas arsine. This gas was then burnt in air to leave a silver–grey crust of elemental arsenic.

An early success for the Marsh test was the prosecution of *l'ange de l'arsenic*, Marie Lafarge. As an orphan in the care of an aunt, the young Frenchwoman was forced into marrying a con artist named Charles Lafarge, who had been found through a marriage agency. Desperate to escape this loveless marriage she set about murdering Lafarge. One of his friends saw her mix a white powder into the sick man's food. She claimed it was gum Arabic, but the Marsh Test showed it to be arsenic trioxide. The test, though too late for Charles, who died the next day, was used as evidence in Marie's murder trial in 1842, which resulted in her life imprisonment. Charles Lafarge was not the only Frenchman sent to the grave by arsenic. It accounted for 33% of all poisonings in France in the 19th century.

o-o

In June 1851 *An Act to Regulate the Sale of Arsenic* criminalized the unrestricted sale of arsenious acid, arsenites, arsenic acid, arsenates and all other colourless poisonous preparations of arsenic in Britain. A vendor had to log the quantity of arsenic sold, the purpose to which it would be put, details of the buyer's profession and their signature. If the seller did not know the purchaser, a witness who knew both had to countersign the transaction. For agricultural purposes the arsenic had to be mixed with at least equal parts soot or indigo to make it distinguishable from sugar. Medical uses were excluded as long as they were formulated according to the prescription of a legally qualified person. The futility of regulating only small quantities was pointed out in 1860 by *The Lancet*: 'To subject men, women, and children to the ingestion of arsenic, by living in an arsenical atmosphere, may at the moment frustrate the aim of justice in the charge of homicidal poisoning'. Later that same year the journal elaborated, in an article entitled 'Arsenic for the millions': 'How to convict arsenical poisoning when ladies use arsenical cosmetics, when confectioners sell arsenical sweatmeats, when paperhangers clothe our walls with arsenical hangings, and impregnate all the air with fine arsenical dust?'.

Perhaps the most famous illustration of this problem was the case of Clare Boothe Luce. Born in New York, Booth Luce married and divorced a millionaire, romanced Randolph Churchill and Vincent Astor, wrote popular plays, worked as an intrepid journalist and was editor of *Vanity Fair.* In 1953 Dwight Eisenhower appointed her Ambassadress to Italy, making her America's first female chief diplomatic representative to a major foreign power. Two years into the post it appeared to be overtaxing her. Her complexion turned sallow, her appetite faltered and her hair was falling out in handfuls. After preliminary medical examinations she was told her symptoms could be explained by anaemia, alcoholism or lead poisoning. None of these seemed to fit, and she was sent to a toxicologist to have her urine tested. It contained arsenic.

The CIA was called in. It appeared that Clare had been receiving small doses of arsenic over some time. First the CIA asked if she and her husband were getting on well, a question she described as 'unforgivable'. The intelligence officers explained that the poisoner had to be someone who had access to her food and drink. Embassy staff immediately fell under suspicion. Infiltration by Italian communists was suspected – she was highly unpopular with them.

The testimony of a maid solved the mystery. A CIA agent noticed an accumulation of a grey dust on an exposed Linguaphone record in Boothe Luce's bedroom. He assumed this was due to poor housekeeping until informed by a maid that she dusted the room – and the record – every morning. The windows were closed and the room was unoccupied all day, yet the following day there was a fresh layer of dust. Samples sent for analysis contained arsenate of lead. It came from the ceiling paint. When someone walked in the room above, or when the American washing machine was used, poisoned dust drifted down from above onto the sleeping Mrs Luce.

The episode sapped Luce's strength so much she resigned her post following year. The arsenic story was kept out of the press for another year. When Eisenhower leaked the story, it appeared under the headline 'ARSENIC AND OLD LUCE'.

o–o

Napoleon Bonaparte was imprisoned by the British on the island of St Helena in the middle of the Atlantic Ocean in 1816. He died there some five years later on 5 May 1821. Autopsy showed that he had stomach cancer, but conspiracy theories abounded, suggesting that the English-organised post mortem was a cover-up. His death has been surrounded by speculation ever since. Did the English kill him, or perhaps plotting French rival camps?

In 1840, to look into the matter further, the French exhumed Napoleon's body from its resting place on St Helena. They got a surprise: the body was extremely well preserved, with bronze skin. Napoleon had not been embalmed, but was hermetically sealed into his coffin. The body was first placed in a tin coffin that was soldered closed. This was set inside a mahogany coffin, which was nested within a lead coffin, again soldered shut. The whole lot was encased in another mahogany coffin with silver screws – a fitting interment for an emperor. Nineteenth century forensic scientists knew from experiments with deliberately poisoned dogs and observations of suicide corpses that arsenic preserves the body from decay. The French investigators opened the coffins for just two minutes and took no tissues for forensic analysis, so as not to desecrate the body.

The story moves forward to 1961, when precious hair mementoes taken from the Emperor before his death were analysed for toxic elements by Hamilton Smith at Glasgow University's Department of Forensic Medicine. The researchers found large quantities of arsenic. Ingested arsenic is quickly incorporated into hair and nails, putting them at the centre of many an investigation.

In their 1982 book *The Murder of Napoleon*, Ben Weider and David Hapgood propose an elaborate explanation for how the Emperor's hair came to contain arsenic. They suggest that Napoleon was given small doses of the poison from his arrival on St Helena in 1816 onwards. They call this the 'cosmetic phase' of the emperor's downfall, and believe that the arsenic was administered via his private wine stock. They implicated Napoleon's wine steward Count Montholon and his French doctor Antommarchi. Weider and Hapgood reckon that this drip-drip approach was not supposed to kill the man, just to make it look like he had a long-term illness.

Next they suggest that during 1821 the poisoning stepped up a gear, to the 'lethal phase', through the administration of toxic medicines. Historical records by those who attended Napoleon at his death detail that he was given the antimony compound tartar emetic and an orange-like drink, orgeat. This last quenches thirst but, being made of oil of bitter almonds, contains hydrocyanic acid (cyanide). Finally, he was given 10 times the normal dose of the laxative calomel – a poison properly called mercuric chloride. This potent cocktail alone could have killed him, intentionally or unintentionally, even without the 'cosmetic phase'.

What complicates the story is that Napoleon's house in St Helena was damp and decorated with arsenic green wallpaper, the perfect conditions for producing trimethyl arsine gas. Antimony in the tartar emetic he took to make him vomit could have contained large quantities of arsenic as an impurity. The prevalence of arsenic in the medicines, cosmetics and hair tonics of the day could also have given the Emperor arsenic-enriched hair. Even dust from a coal fire could have played a part.

The mystery may have an even more prosaic explanation. Victorians often sprinkled mementoes with arsenic to stop them decomposing. I have tested hair samples preserved in sentimental jewellery from the early 19th century. Some have enormous levels of the stuff. Conspiracy theories make good reading, but from a century awash with arsenic, arsenic in the hair means little.

In 2004 the notion that Napoleon was deliberately poisoned evaporated, for all but the most ardent supporters, when Xilei Lin and co-workers at Munich university analysed locks of hair taken before his exile on St Helena. Those from 1814 and 1815 had the highest arsenic content. Toxicologists agree that if the arsenic in Napoleon's hair was coming from his food or wine, he should have been dead well before 1821.

Probably the most fanciful rationalization of all this came from one Charles Boner in *Chambers Journal* in 1856. Boner was convinced that Napoleon, like the legendary Mithridates, took arsenic to build up his immunity to poisoning.

o-o

The ubiquity of arsenic as a medicine also rendered it the ideal agent for suicide. This trend probably inspired Gustave Flaubert's *Madame Bovary*, published in 1856: 'She lifted down the blue jar, pulled out the stopper, rummaged inside with her had, and withdrew it filled with a white powder. This she began to eat on the spot'.

In a letter dated November 1866 Flaubert described killing off his heroine: 'When I wrote the description of the poisoning of Mme Bovary I had the taste of arsenic so much in my mouth, I had taken so much poison myself that I gave myself two bouts of indigestion one after the other – two real bouts for I threw up all my dinner'.

Flaubert seems to have made one mistake. While looking up the symptoms of arsenic poisoning in a medical book it appears he read the mercury section on the same page. 'The disgusting taste of ink' that his Bovary experiences is characteristic of mercury, not arsenic. He mistook the inky taste in his own mouth for a manifestation of his imaginative immersion. It was more likely a reaction to the mercury he was taking for syphilis.

o-o

Back to homicide, and an inferior English rip-off of *Madame Bovary*, Mary Elizabeth Braddon's *The Doctor's Wife*. Braddon was extraordinarily prolific, producing over 80 'sensation' novels. In this one, from 1864, she tampered a bit with Flaubert's plot, having her frustrated heroine instead contemplating murdering her doctor husband with his own arsenic supply:

> ...the possibility of deliberately leaving her husband to follow the footsteps of this other man, was as far beyond her power of comprehension as the possibility that she might

> steal a handful of arsenic out of one of the earthenware jars in the surgery, and mix it with sugar that sweetened George Gilbert's matutinal coffee.

Indeed, arsenic poisoning has become a staple of crime writing, well used by Agatha Christie and Dorothy L. Sayers. In Sayers' murder mystery *Strong Poison*, the guilty party eats an arsenic-spiked meal with his victim but survives. His lustrous hair and excellent complexion give him away to Sayers' dapper detective, Lord Peter Wimsey, as an *arsenic eater*:

> ...about this arsenic. As you know, it's not good for people in a general way, but there are some people – those tiresome peasants in Styria one hears so much about – who are supposed to eat it for fun...

Sayers was probably inspired by the rise of the 'Styrian defence' in late 19th century litigation. This curious defence was used when real-life Southern belle Florence Maybrick was tried for the murder of her English husband in Liverpool in 1889. Florence and her dead spouse had a serious falling out when she started seeing another man. The corpse contained large quantities of arsenic. Florence claimed that her husband took arsenic regularly for his health – part one of the Styrian defence. Forensic scientists found arsenic in his medicines and on a jug that Florence had used to prepare his food.

They also found a stash of white arsenic and arsenic flypapers in Mrs Maybrick's bedroom. (Soaked flypaper was known to yield up to 400 mg of arsenic in a brown solution, often disguised by poisoners as strong tea, Bovril or brandy.) Florence claimed she was using the solution for her skin – part two of her Styrian defence. At her trial she said:

> The flypapers were bought with the intention of using as a cosmetic. Before my marriage, and since, for many years, I

> have been in the habit of using a face-wash prescribed by Dr. Greggs, of Brooklyn. It consisted principally of arsenic, tincture of benzoin, elderflower water, and some other ingredients. This prescription I lost last April, and as at that time I was suffering from slight eruption of the face, I thought I should like to try to make a substitute myself. I was anxious to get rid of this eruption before I went to a ball on the 30th of that month.

Maybrick's Styrian defence failed: Florence was sentenced to life imprisonment. As a result of public protest she was eventually released in 1904, a year of great fictional note. The 16th of June 1904, now commemorated yearly as Bloomsday, is the date on which James Joyce's Leopold Bloom wandered Dublin in *Ulysses* (published in 1922). Joyce wove Maybrick into Bloom's stream of consciousness:

> ...take that Mrs Maybrick that poisoned her husband for what I wonder in love with some other man... white Arsenic she put in his tea off flypaper wasn't it...

o-o

The Styrian defence was more successfully deployed during the Edinburgh trial of Madeleine Smith, charged with the murder of her lover in 1857.

Madeleine was the daughter of a wealthy Glasgow architect. Aged 21, she was accused of poisoning a poor shipping clerk, Pierre Emile L'Angelier, her secret fiancé. The trial became a national sensation when her letters to him, read out in court, revealed that they had indulged in premarital sex, taboo in polite Victorian society.

Madeleine was 19 when she began to have liaisons with the 26-year-old L'Angelier, a Jersey man of French extraction. He had been sent to Scotland in 1842 to gain experience for his family's seed business. He left Edinburgh in 1852 for Glasgow, taking up the job of a shipping clerk in a trading firm. Despite being a hard-working churchgoer, Emile was restless and dissatisfied, spending most of his meagre wages on stylish clothes. Parading around town one day he met Madeleine. A clandestine romance ensued, owing to their gulf in status.

During their affair Madeleine found a partner of more suitable social standing from whom she accepted a marriage proposal. When Emile discovered this he blackmailed her, threatening to show her father her scandalous love letters. Shortly afterwards, on 22 March 1857, Emile died suddenly after three bouts of sickness, coinciding with Madeleine's purchase of arsenic trioxide from a pharmacist. Emile told a confidant that he had received chocolate and coffee from Madeleine just before his illnesses, and wondered if she was poisoning him, as she had good reason to.

The post mortem revealed arsenic poisoning. The police obtained Madeleine's letters and discovered her recent arsenic purchases. She was arrested and charged with murder.

In court, Madeleine maintained she had told the pharmacist that the arsenic was for poisoning rats, because she was too embarrassed to say she wanted it for a skin wash, having heard of its benefits at school. This was part of the Styrian defence. The other was that Emile, known to be a dandy, dressing well above his station, could have been vain enough to be an arsenic eater. Madeleine's lawyers suggested that he could have picked up the habit early in life when he worked at a stable – feeding horses arsenic was common at the time.

Though probably guilty, Ms Smith was acquitted on a 'not proven' verdict, a peculiarity of Scottish law, helped considerably by her class and the public's distaste for the blackmail of a young lady.

Her release was greeted with huge support by some newspapers. She was the victim of a seducer and treasure hunter, they agreed, and if she did poison him, it was righteous revenge. Others were less inclined to Miss Smith's cause, including the *Glasgow Sentinel:*

> [Madeleine Smith was] as much the seducer as the seduced. And when once the veil of modesty was thrown aside, from the first a very frail and flimsy one, the women of strong passion and libidinous tendencies at once reveals herself.

The *Examiner* had a similar impression:

> To Madeleine Smith alone his horrible death seems to have been no shock, no grief, and she demeaned herself [at] her trial as if L'Angelier had never had a place in her affections. If it had been a trial for poisoning of a dog the indifference could not have been greater.

Madeleine's dealings bring to mind a passage from Alexandre Dumas's *Count of Monte Cristo* (1844–45):

> Amongst us a simpleton, possessed by the demon of hate or cupidity, who has an enemy to destroy, or some near relation to dispose of, goes straight to the grocer's or druggist's, gives a false name, which leads more easily to his detection than his real one, and under the pretext that the rats prevent him from sleeping, purchases five or six grammes of arsenic – if he is really a cunning fellow, he goes to five or six different druggists or grocers, and thereby becomes only five or six times more easily traced; – then, when he has acquired his specific, he administers duly to his enemy, or near kinsman, a dose of arsenic which would make a mammoth or mastodon burst, and which, without rhyme or reason, makes his victim utter groans which alarm the entire neighbourhood. Then arrive a crowd of policemen and constables. They fetch a doctor, who opens the dead body, and collects from the entrails and stomach a quantity of arsenic in a spoon. Next day a hundred newspapers relate the fact, with the names of the victim and the murderer. The same evening the grocer or grocers, druggist or

> druggists, come and say, 'It was I who sold the arsenic to the gentleman;' and rather than not recognize the guilty purchaser, they will recognize twenty. Then the foolish criminal is taken, imprisoned, interrogated, confronted, confounded, condemned, and cut off by hemp or steel; or if she be a woman of any consideration, they lock her up for life.

To escape her notoriety Madeleine Smith fled Glasgow society. But arsenic was to touch her life again. She joined the entourage of William Morris.

o–o

Madeleine met one George Wardle. Wardle was a draftsman and antiquarian, and friends with Philip Webb (one of the founding members of Morris, Marshall, Faulkner & Co.). George was attracted to Madeleine, and ignored warnings about her past with pre-Raphaelite gallantry. He proposed, she accepted, and they moved to London after their marriage on 4 July 1861.

Because of George's job and contacts, Madeleine was quickly accepted in the *avant garde* London set. George became William Morris's business manager and Madeleine got to know William's wife Jane, and was soon embroidering for the family. Violet Hunt, a friend of Morris's daughter May, recollected a visit to the Morrises at Kelmscott House: 'Mrs Morris was famous for her embroidery, and sitting "lily-like arrow", with Red Lion Mary [Morris's house keeper] and Mrs W. [Wardle], who was really Madeline Smith, laid with her needle the foundations of the firm of Morris, Marshall and Faulkner & Co.'.

Madeleine's social set probably knew of her former life. Jane Morris certainly did, as revealed in a rather nasty play sent to her by her lover, the painter and poet Dante Gabriel Rossetti, shortly after Morris dissolved his business in 1875, much to Rossetti's disgust. The play has George and Lena (Madeleine) poisoning Morris in order to inherit Morris, Marshall, Faulkner & Co.

Madeleine became active in the London socialist movement centred on Morris's circle. George Bernard Shaw described her as 'an ordinary good-humoured capable woman with nothing sinister about her'. He recorded that she cooked dinner for a political club in the East End until the leading socialist Belfort Bax discovered her identity and went around telling members they were in danger of being poisoned.

Geoffrey Butler's 1935 biography of Madeleine tells of a party around the time of the Florence Maybrick trial. The journalist George R. Simms, supping with the Wardles, started to discuss criminology and Madeleine Smith's name came up. He remarked that if it was true, as was reported, that she was alive and married, 'her husband must be a brave man'. His neighbour whispered in his ear that Madeleine Smith was at the table with her spouse.

Other literary notables commented on Miss Smith. George Eliot said 'I think Madeleine Smith one of the least fascinating of murderesses, and since she is acquitted it is a pity Palmer is not alive to marry her and be the victim of her second experiment in cosmetics – which is too likely to come one day or other'. Henry James, on the other hand, was captivated. As a boy of 14 he followed the trial in *The Times*, delivered to his parents' home in Bologna. In 1914 he wrote: 'And what a pity she was almost of the pre-photographic age – I would give so much for a veracious portrait of her then face'. He clearly never saw Figure 6.2.

Madeleine's marriage to George yielded two children, but did not last. In straitened financial circumstances she moved to Leek in Staffordshire, the home town of her brother-in law Thomas Wardle, another Morris connection. Thomas was a silk dyer who remembered natural dyeing processes from his youth. It was to Thomas that Morris turned to learn dye craft, spending much time in Leek itself.

When Thomas died the 70-year-old Madeleine moved to the USA where her son had settled. There she died in anonymity aged 93.

Figure 6.2 A rare photo of Madeleine Smith that Henry James missed.

⚯

Arsenic poisoners are not a thing of the past. This quote is from *The Independent*, 28 May 2003, in article entitled 'Church of the poisoned mind', concerning a fatal religious service on 28 April 2003 in the tiny community of New Sweden, Maine, USA:

> For Eric Margeson, the first sign that something might be wrong with the coffee the church members had prepared at the end of the service was the aftertaste: it was too bitter and, as he swallowed, it burned the back of his mouth. Within minutes he was reeling, suddenly feeling nauseous. Back at his father's house, just a few miles from the church, he could not stop throwing up.

Eric was one of 16 people poisoned with coffee – one, Walter Morrill, died. Initially Daniel Bondeson, a member of the congregation, was the chief suspect, as he fatally shot himself five days after the poisoning, and was found with an incriminating note at his side. But three weeks after the poisoning the police investigators thought that someone else was also involved, and had narrowed it down to

10 people. They believed that the motive lay in 'the dynamics of the church'.

The mystery is yet to be solved. Where the arsenic came from is clear, however. In this farming community the poison is regularly used to defoliate potato crops. There was plenty 'tucked away and forgotten on a shelf in one of the vast wooden barns that dot the landscape'.

Masumi Hayashi, the most notorious of recent mass arsenic poisoners, is appealing against the death penalty as I write. She has been convicted of spiking a curry served at a summer festival in Wakayama City, Japan. Two children and two adults died; 63 others survived. The curry contained enough arsenic to kill up to 1,350 people. Masumi refused to speak at the trial, but her criminal husband ran a termite extermination business using arsenic as an insecticide. And traces of the arsenic found in her kitchen were linked to that in the curry after analysis in a synchrotron: a huge particle accelerator.

Chapter 7
NOTHING GREEN MET THE EYE

There was a lord that high Maltê,
Among great lords he was right great,
On poor folk trod he like the dirt,
None but God might do him hurt.
Deus est Deus pauperum

William Morris, *The God of the Poor*, published in the *Fortnightly Review*, 1868

Before 1800 nearly all valuable arsenic products came from Germany, particularly the mines of Saxony. Arsenic was a by-product of roasting a cobalt ore for the production of zaffre – a cobalt compound used to colour ceramic glazes blue (Figure 7.1). The corrosive effect of these ores was recorded by Georgius Agricola in his seminal work of 1556 on mining and mineralogy *de Re Metallica* (incidentally translated into English by American President Herbert Clark Hoover). Agricola describes the arsenic–cobalt compound smalite ($CoAs_2$), which he calls cadmia: 'Further, there is a certain kind of cadmia which eats away the feet of the workmen when they become wet, and similarly their hands, and injures their lungs and eyes'. Saxony's miners thought that a goblin in this ore accounted for its unreasonable behaviour.

It was the south-west of England, however, that supplied the vast amounts of arsenic consumed by Victorian enterprise. Base and precious metal mining was a way of life for the inhabitants of

Figure 7.1 Arsenic production from cobalt ores, from a 1768 edition of Dennis Diderot's *D'Alembert Encyclopedie.*

the counties of Devon and Cornwall. Cornwall's pre-eminence in tin supply stretched from Roman times until the beginning of the 19th century, when Malayan ore fields were opened. Fortunately for the mining fraternity, exploration of Cornish and Devonian mineral districts at the same time revealed massive copper deposits.

In the days before major industrial outlets were found for arsenic, its co-occurrence with copper, tin, silver and lead, as what miners called 'mundic' (arsenopyrite), was a considerable nuisance. It makes processed tin and copper brittle. So women and girls, known as 'bal maidens', were employed to hand-sort arsenic minerals from tin and copper ores before smelting. Mounds of discarded arsenopyrite dotted the countryside. This waste was often used to surface roads.

After sorting, tin ore was roasted or 'calcined' in facilities termed 'burning houses' to drive off any remaining arsenic prior to sale to the smelter. Ore was loaded into a furnace and raked over until it ceased to give off white arsenic fumes. These fumes gathered inside the burning house chimneys or were released into the surrounding countryside as a poisonous fog.

The environmental impacts of copper and tin smelting were grim. Visitors to mining districts were astonished to see the vegetation on the upper slopes of the valleys bathed in arsenic fumes. The Plymouth poet N.T. Carrington summarised the impact of the mining boom on the landscape in his 1828 poem *The Banks of the Tamar*:

And seas of vendure roll'd, vile rubbish meets
The eye disgusted. How the thirst of gain
Drags from the bowels of our mother Earth,
The unsightly masses, strewing there once
Wav'd the gay woods, or babbled the soft brooks!
And where the blossoming orchards bless'd the view
Tremendous arsenic its fatal fumes
Has breath'd and vegetative life has ceas'd

And desolation reigns. With dauntless hand
Deep into realms of gloom the miner cuts
His desperate way. Mephitic vapours hang
Around the bold adventurer, and pale
Consumption threatens; still he works, and rears
His subterranean galleries, and oft,
Pursuing the very jaws of Death
The all-attractive lode, he hears above.
Appalling sound – the eternal Ocean howl!

John Ayrton Paris practised medicine in Cornwall, where he experienced copper and tin smelting facilities at first hand. In his widely read book *Pharmacologia,* published in a number of editions through the 1820s, he wrote:

> ...it may however be interesting and useful to record an account of a pernicious influence of arsenical fumes upon organised beings as I have been enabled to ascertain in copper smelting works, and tin burning houses of Cornwall. This influence is very apparent in the condition of both animals and vegetables in the vicinity, horses and cows commonly looses their hoofs, the later are often to be seen in the neighbouring pastures crawling on their knees and not unfrequently suffering from a cancerous affliction in their rumps, whilst the milk cows, in addition to these miseries, are soon deprived of their milk; the men employed in the works are more healthy than we could *a priori* have supposed possible;.... It deserves notice that the smelters are occasionally affected with a cancerous disease in the scrotum. Similar to that which infests chimney-sweepers.

Laxness in controlling arsenic pollution is explained by Paris in his book *Medical Jurisprudence:* '...there are circumstances which ought to exempt certain establishments from the operation of common law; we allude to those grand national works for smelting ores, which could not be closed without totally affecting our national prosperity'.

Arsenic-related disease was first noticed in the city of Reichenstein in Silesia, a gold and silver mining region for nearly a millennium. Contamination of drinking water supplies meant that the inhabitants suffered chronic arsenic poisoning, with symptoms such as liver, skin, gastrointestinal and nervous system disorders termed 'Reichenstein disease'.

o–o

The beginning of organised arsenic production in Britain came in 1812 when a tin smelting house at Perranarworthal, Cornwall, was adapted for arsenic trioxide manufacture: 'and it is able to manufacture a sufficiency for the supply of this Kingdom and of a quality fully equal to the best'.

Demand for arsenic was grew, and a new plant was opened in 1833 or 1834 in nearby Bissoe. Around 1840, one Thomas Garland of Redruth, who had acquired by then the original Perranarworthal works, decided to produce arsenic from the mineral arsenopyrite in addition to refining the crude arsenic brought from the tin-burning houses of Cornwall. Tens of thousands of tons of this material was lying around Cornwall, available for just the price of carting it away. Roads covered with arsenic minerals were torn up to meet demand.

From the arsenic factories fumes poured out night and day. In 1851 Garland was prosecuted 'for creating and continuing a public nuisance' when one Richard Thomas, who was forced to leave his home 200 metres from the Perranarworthal plant, took him to court. The smoke had killed a cow, two heifers, two horses and a pony grazing in Thomas' fields. His fruit trees were blackened and ruined, and he was unable to grow vegetables, except potatoes which were immune to the arsenic. Thomas complained that he had to shut all his doors and windows and could not dry his washing outside 'when the wind blew from a certain quarter'.

Figure 7.2 Arsenic condensed in a chamber.
From *Scientific American*, 1907. Reproduced with permission.

o–o

Arsenic refineries used a kiln, as for tin. The outlet passed into flues 100 to 200 metres long, cooling the gases and causing arsenic trioxide to condense in chambers situated along the flues (Figure 7.2). Very pure arsenic trioxide collected in chambers nearest to the furnace. Further along, a white arsenic–sulphur mix gathered. At the furthest reaches, mainly sulphur condensed.

The sulphur was a valuable by-product, used to make sulphuric acid. The English had depended on sulphur from volcanoes in Sicily, but in 1838 a French company held a monopoly on this source and prices rose. Cornish sulphur provided an alternative. The boom lasted until 1856, when Sicilian sulphur was imported once more and Spanish sulphur began to come in from Huelva's Rio Tinto mines, where copper ores were richer in the element than those in Cornwall. Cornish sulphur, being of lower quality because of its arsenic contamination, fell in price.

o–o

In its heyday, copper was too profitable for mine owners to be concerned with capitalising on its polluting arsenic by-product. But when copper prices crashed in the 1860s industrialists realised there was money to be made from selling arsenic for the production of glass, enamel, leather, fireworks, pigments and lead shot (it kept the shot spherical during manufacture).

One copper mine produced between a third and a half of the world's arsenic during the 19th century. Called Devon Great Consols, the mine, situated in Devonshire's Tamar Valley on the border with Cornwall, opened in 1844. In its arsenic-producing prime, Devon Great Consols boasted among its Board of Directors our old friend William Morris. In 1899, Mackail, Morris's first biographer, described Devon Great Consols as follows:

> It leads a somewhat struggling existence on the proceeds of arsenic which, in the high days of the copper-mining industry, was neglected as an unimportant by-product. But its earlier fortunes, and its gradual decline, were not without importance in determining the course of Morris's life.

To put it more bluntly, Devon Great Consols made William, and the rest of his family, fabulously rich. It was this wealth that launched the greatest design firm of the 19th century, Morris & Co.

Morris never saw the source of his wealth, although he came tantalisingly close. In a letter to his wife Jane, he describes an idyllic February day in 1881 visiting his sister Alice Gill who lived in south Devon near the Cornish border. 'The day was bright after the morning down there & Alice took me a beautiful walk to the Tavy near where it runs into the Tamar'. Devon Great Consols lay just eight kilometres due north.

Being the greatest and richest copper mine in Britain, Devon Great Consols became renowned as a magnet for speculators. Indeed, so famous was the mine that it entered into literature, albeit thinly disguised. 'It will be sufficient for us to know that

Wheal Mary Jane was at the moment the richest of all the rich mines that had opened in that district' wrote Anthony Trollope in his 1858 novel *The Three Clerks*, depicting the greed of the get-rich-quick share dealing of the day. The fictional Wheal Mary Jane is located just outside Tavistock, a real old stannery (tin) town a couple of miles from Devon Great Consols.

Trollope, who despised the human and environmental cost of mineral mining and hated its money-grabbing entrepreneurs, immortalised the activities at Devon Great Consols as follows:

> It was an ugly uninviting place to look at, with but few visible signs of wealth. The earth, which had been burrowed out by these human rabbits in their search after tin, lay around in huge ungainly heaps; the over ground buildings of the establishment consisted of a few ill-arranged sheds, already apparently in a state of decadence; dirt and slush, and ponds of water confined by muddy dams, abounded on every side; muddy men, with muddy carts and muddy horses, slowly crawled hither and thither, apparently with no object, and evidently indifferent as to whom they might overset in their course. The inferior men seemed to show no respect to those above them, and the superiors to exercise no authority over those below them. There was, a sullen equality among them all. On the ground around was no vegetation; nothing green met the eye, some few stunted bushes appeared here and there, nearly smothered by heaped-up mud, but they had about them none of the attractiveness of foliage. The whole scene, though consisting of earth alone, was unearthly, and looked as though the devil had walked over the place with hot hoofs, and then raked it with a huge rake.

Did Morris know that the beautiful Devon countryside he enjoyed that bright winter's day would not be found at the mines? Was he aware that his wealth came at the price of despoiling the once-pristine Tamar Valley with massive heaps of toxic slag and poisonous fumes? We will never know. Certainly by 1881 his writings railed against the environmental devastation wrought by the Victorians' pursuit of wealth. But no record survives of him referring to – let alone showing remorse for – his own role in this dirty business.

Even today visitors looking left as they cross into Devon from Cornwall across the little bridge at the mining village of Gunnislake are confronted with a scene not too dissimilar from that which Trollope described. Though now disused, the barren, scarred Devon Great Consols site is still devoid of vegetation – a sad reminder of the financial reality responsible for William Morris's beautiful legacy.

o-o

Morris was born into commerce. His father, William Morris senior, was a successful discount broker. On commission, he traded orders to pay a fixed amount of money on a fixed date to raise money to fund business ventures. These orders were important to the British economy of late the 18th and early 19th centuries, as overdraft and loan facilities were limited. Discount brokering was a risky business, and those that were good at it, and could survive market fluctuations, became wealthy. Family ties secured William Morris Snr a post in a prosperous discounting firm, Sanderson & Co. In 1826, the 28-year-old Morris Snr was made partner. As the firm prospered he married Emma Shelton. They had two girls, Emma and Henrietta, and, in 1834, William Jnr.

The family moved to the spacious Elm House in Walthamstow, today in the suburbs of North London, then a leafy village. As money poured into Sanderson & Co., the Morris family moved to a large Georgian mansion in the Palladian Style, Woodford Hall. The house sat in 50 acres of parkland, surrounded by Epping Forest and farms.

To mark his rise through society, Morris Snr obtained a grant of arms from the College of Heralds. The family had arrived, and lived very well. But Morris Snr gambled for more, exploiting his business connections. He got into the flourishing shares market of the mid-1840s satirised in *The Three Clerks*. With his brothers Thomas and Francis, both in the coal trade, he speculated in copper mining ventures in the west of England. Copper prices were high and there were fortunes to be made.

Most of the copper speculations were unsuccessful, but the possibility of hitting a large lode kept investors queuing up. The man the Morrises gambled on, Josiah Hitchens, had a track record of success, having already found the third biggest copper seam in Devon. Hitchens heard that local miners suspected there to be a large copper seam in Blanchdown Wood, a pheasant wood owned by the Duke of Bedford. Although the Duke had declined to give previous speculators access to his wood to 'maintain the privacy of his pheasants', he was impressed by Hitchens and relinquished his hunting to the mining concern. He reaped considerable reward for the loss of his land.

The Duke of Bedford was no conservationist, reluctant to sacrifice nature to commerce. He loved his hunting so much that the 'curse of the pheasant' was on the Tamar Valley. Gamekeepers systematically killed sparrow hawks, buzzards, weasels and badgers from Blanchdown Wood. Locals were forbidden to walk across the estates in case they disturbed the birds. All this cosseting ensured that a vast slaughter could be had in the two or three big shoots of the season.

Hitchens, having secured the Duke's blessing, went to London to raise the capital for his venture. William Snr and Thomas Morris bought shares in the Devonshire Great Consolidated Copper Mining Company – Devon Great Consols for short.

On 4 November 1844, when William Morris junior was 10 years old, the exploratory party of miners working a disused shaft in Blanchdown Wood hit what was then the greatest copper lode in the world.

Thomas and William Snr. were appointed directors of the company. When it was apparent that the mine was going to be successful and open for some time, Thomas moved from London to Tavistock in Devon to serve as the company's resident director. He settled down to building Abbotsfield Hall in 1852.

o-o

Devon Great Consols was registered as a joint stock company with 1,024 shares. Hitchens held 144 ; Richard Gard, the MP for Exeter and a London discount broker, held 288. Two city stockbrokers, William Thomas and John Thomas held 144 shares each. William Morris Snr held 272 shares. The remaining 32 belonged to his brother Thomas, who with Gard and William Thomas formed the initial board of directors. The board was enlarged in May 1846 to include William Morris Snr and John Thomas.

After Devon Great Consols' first year of trading an original £1 share changed hands for £800, with the first year's return on the share dividend being £71. A year later the shares slumped to £155, with the return at £25. This fall was due not to declining returns but the need for more infrastructure to support the massive scale of the operation. Large pumps, railways and quays all had to be constructed. Between 1850 and 1856 dividends averaged £49 per share. Devon Great Consols eventually paid its investors dividends of £1,200,000 for an initial stake of just £1,024.

The pound today is worth approximately 80 times its 19th century value. Thus in today's money, £80 a share increased to £64,000 after one year, and William Morris senior's investment of £21,760 pounds rose to £17.4 million. The total dividend over the entire life of the mine from the 1,024 shares would today be worth £96 million. No wonder that Devon Great Consols became such a symbol of the wealth that the stock market could bring.

o–o

Three years after Devon Great Consols opened, the Morris family was hit with tragedy. William Morris Snr died suddenly, aged 50. Seven days after Morris's death his company, Sanderson & Co., ceased trading with debts totalling £2,606,569. The failure of the business had its roots in the policies of the Bank of England. The Bank Act of 1844 allowed it to lend and discount freely at low rates of interest, competing with the discount houses for business. In 1847 it reversed this policy due to plummeting reserves, causing the discount houses considerable problems. The railway investment boom of 1844–47 was quickly checked, and the discount houses came under pressure from creditors concerned at the value of their bills. Sanderson's situation was even more desperate, as it held a lot of corn bills, and the price of corn also fell. With the death of Morris, customers lost their nerve.

In the long run Sanderson's remained solvent as its creditors numbered leading London bankers who expressed confidence in the firm. In the short term, however, the collapse of Sanderson's had considerable financial consequences for the Morris family. They had lost William Morris Snr's wage and his share of the partnership's capital. Mrs Emma Morris probably had to liquidate of some their personal assets to meet creditors despite her husband's estate being valued at £60,000 – with the 272 shares in Devon Great Consols accounting for an estimated two-thirds of this.

The family could no longer afford Woodford Hall and moved to Water House, a smaller home in Walthamstow that today houses the William Morris Museum. Devon Great Consols was now the family's lifeline. This was a precarious situation as copper prices rose and fell. Nonetheless, the Morris's income over the first 25 years was very healthy. When the share price rose in 1850, Emma sold 72 to raise over £20,000, which she reinvested in safer, less profitable securities. Still, the annual dividend from the Devon Great Consols shares kept Emma Morris and her children in great comfort.

Following the death of William Snr his brothers Thomas and Francis became Emma's financial advisors. Francis bought 20 shares from her in 1849 and replaced his dead brother on the mine's board of directors. The brothers advised Emma to divide 117 shares between her nine children, 13 each when they reached maturity, to give them a good start in life.

Traditionally mines are named after the wives of investors and owners. So as a tribute to William Morris senior, a year after his death, one of the shafts at Devon Great Consols was named Wheal Emma – 'wheal' being Cornish for mine.

⚯

All this wealth gave William Morris Jnr the perfect environment to refine his remarkably broad talents, eventually developing into one of the greatest artists, designers, poets and novelists of the 19th century. In 1955 the historian E.P. Thompson, in his communist apologist biography, pointed out the dichotomy of William's early life compared with his later revolutionary socialism.

> The handsome dividends came regularly into the rural village of Walthamstow, on the edge of Epping Forest, bringing with them nothing to indicate the miseries at the bottom of the cramped and ill-ventilated shafts from which they had their source.

In 1864 the suffering of the Devon copper miners was illustrated in a landmark study by a Royal Commission under the chairmanship of the reformer George William Kinnaird. The ninth Baron Kinnaird toured England and Wales investigating the social and labour conditions of miners. It was the first coordinated survey of working life, health and accidents in Britain's copper, tin and lead mines, and was conducted at the peak of their fortunes. The report and its suggested reforms were badly received by the management and shareholders of the mines, always concerned with profit margins.

The Royal Commission recorded over 23,000 questions and answers verbatim from 250 witnesses. One witness was William Morris's uncle, Thomas Morris, the resident director of Devon Great Consols. Thomas not only saw the poverty and suffering of the miners at first hand, he was directly responsible for it. Practices at Devon Great Consols raised particular concern with the Commission according to the transcript of Thomas Morris's questioning by Kinnaird on 26 May 1863.

Morris admitted that young girls were employed at the age of eight sorting ore at the surface for 4d per day (around £1.30 today). Boys started down the mines as young as 10 and were on the scrapheap by 30, exhausted by dust, explosives and daily climbs of 400 metres up and down which left them with chills and heart problems.

The Commission also questioned the Tavistock surgeon, who looked after the workers of Devon Great Consols. Each family paid 9d a month to stay on his books, with the money deducted from miners' pay. If men could not work due to ill health they stopped having their medical bills paid, one of the major concerns of the Commission.

The surgeon reported that measles, scarlet fever and cholera were rife due to his patients 'living in impure air in consequence of the houses being densely crowded'. He observed 'miners disease', which the men got around 40, starting with black spit, developing into bronchitis, and terminating in consumption. Today we know this as silicosis. The doctor argued that his clients' living conditions caused the silicosis, and was keen not to blame the mines, his source of income. However, when asked if miners' wives got the disease he said no.

The medical report of the Kinnaird Commission, presented by James Bankart, noted that tin and copper miners were:

> ...as a class, pale, thin, and sallow. This appearance is rendered more striking by contrast with the women and

> children, who are fresh coloured, stout, well developed, and every respect a healthy looking class.... The unhealthy look of the men increases in proportion to their years, to the length of time they have been employed underground, and to the nature of their work during that time, and is especially evident in men beyond about 35 years of age.

The report was discussed in newspapers and its findings were common knowledge.

Around this time William Morris was making his name. He established Morris, Marshall, Faulkner & Co. in 1861, and produced his first poisonous wallpapers in 1864. Later the arsenic for the green dyes he deployed in patterns such as *Trellis* and *Indian* probably came from Devon and Cornwall. The Firm, as it was known to the partners, was central to the 19th century Arts and Crafts revival that spurned mass-produced goods in favour of hand-made items created using medieval techniques.

o-o

By 1867 lower grade ore accumulating around the surface of Devon Great Consols was becoming a problem. So the board decided to use it to produce arsenic. The £2000 works were completed and producing 50 tonnes per month by December 1868, when it was decided to enlarge them to produce 150 tonnes a month.

The works consisted of furnaces where intense heat vaporised the arsenic in the ore. Vapour was then conducted through baffles in which it condensed to a greyish-white powder. Excess vapour was then passed through a flue, often stacked with brushwood to trap remaining arsenic, before being discharged from a large chimney. Baffles in the condensing chambers were opened after several weeks and the arsenic dug out by hand. Further refining took place in a furnace in which anthracite purified the crude arsenic. It was then condensed as small crystals of pure white arsenic, which were ground into a fine powder.

The arsenic works at Devon Great Consols were the largest in Devon and Cornwall, and indeed the world, during the 19th century, covering eight acres, with five ovens, three refineries and 1,200 metres of flues. This setup produced 2,500 tonnes of arsenic a year in 1871, increasing to 3,000 tonnes in 1884 and 3,500 tonnes in 1891.

o-o

In 1871, a decade before his conversion to Marxism, the 37-year-old William Morris became a director of Devon Great Consols. He celebrated this appointment by purchasing a top hat for wearing to directors' meetings.

When he joined the board the glory days of Devon Great Consols were over, with the copper lode nearing exhaustion and cheap imported metal from America, Australia and Spain causing prices to plummet.

William knew that arsenic was vital to Devon Great Consols, and hence his family's bank balance. He wrote to his mother in November 1872 following a meeting of the board of directors: 'Last Friday we had the new contract for arsenic, and got a very good price for it'.

From this point on, the fortunes of Devon Great Consols went into terminal decline. Rising exploration costs and slumping copper prices meant that by 1874 Morris started offloading his shares, selling 80 of them for only £80. He sunk his energies into Morris & Co., of which he became the sole proprietor of in 1875.

In the same year he resigned his directorship of Devon Great Consols, symbolically sitting on his top hat. He was becoming increasing interested in socialism, and the move severed his links with Victorian capitalism whilst being highly financially astute.

Devon Great Consols was also starting to draw some very bad press. The quantities of arsenic it was producing were vast. *The Times* began to ask questions. There was arsenic on the site 'probably sufficient to destroy every living animal on the face of the earth' an 1875 article quailed. 'There is at present no efficient law to prevent many-fold this amount of this deadly material from being cast every month into the rivers and watercourses of this country', the piece went on. A quite justified fear, as it turned out: the run-off from Devon Great Consols destroyed the salmon fishery of the River Tamar.

J.T. Arlidge in his *The Hygiene, Diseases and Mortality of Occupations*, 1882, ruminated on the arsenic production of Devon Great Consols: 'The question forces itself upon the mind – where does all this poisonous substance go, and for what legitimate purposes is it employed? I have not sufficient information to supply an answer, but I partake of the surprise most people must feel who hear of so enormous a production of the poison'.

Nonetheless, the Morris family ensured that its strong links with Devon Great Consols remained on William's resignation. His brother, Hugh Stanley, a Hampshire gentleman farmer, took his place on the Board of Directors.

o–o

When Morris became a leader of the British socialist movement in the 1880s he raged at the destruction wrought by capitalists, as illustrated by a lecture given on 14 November 1883 at the Russell Club, a Liberal Organisation at Oxford, in which he said:

> To keep the air pure and the rivers clean, to take some pains to keep the meadows and tillage as pleasant as reasonable use will allow them to be.... Surely not an unreasonable asking. But not a whit of it shall we get under the present system of society.

Or as he put it in his essay *Commerce is an evil*:

> ...but there are other matters not merely useless, but actively destructive and poisonous which command a good price in the market; for instance, adulterated food and

> drink. Vast is the number of slaves whom competitive Commerce employs in turning out infamies such as these, But quite apart from them there is an enormous mass of labour which is just merely wasted; many thousands of men and women making nothing with terrible and inhuman toil which deadens the soul and shortens mere animal life itself.

And again an excerpt from his lecture *Against the Age*:

> Cut down the pleasant trees among the houses, pull down the ancient and venerable buildings for the money that a few square yards of London dirt will fetch: blacken rivers, hide the sun and poison the air with smoke and worse, and it is nobody's business to see it or mend it: that is all that modern commerce, the counting-house forgetful of the workshop, will do for us herein.

It is for utterances like these that Morris is viewed as a founding father of the eco-left movement. His anger about the pollution caused by industrialised humanity was real. How much was he concerned that the foundation of his personal wealth was the most poisonous and dangerous of industries, copper and arsenic mining and smelting? Search Morris's writings and there is little to answer this question. There is no word on the environmental and human damage done by Devon Great Consols in his published works, and as far as we can tell from his extant correspondence, there was none in his private exchanges.

Only one note of regret survives of Morris's involvement with the greatest mineral mining concern in 19th century Britain. In his sentimental utopian dream *News from Nowhere*, published in 1890, Old Hammond, Morris's guide to the historical changes brought about by the socialist revolution in his futuristic Britain, describes to the time-traveller at the centre of the novel changes in mineral mining practices:

> For the rest, whatever coal or mineral we need is brought to grass and sent whither it is needed with as little as possible dirt, confusion, and distressing of quiet people's lives. One is tempted to believe from what one has read of the condition of those districts in the nineteenth century, that those who had them under their power worried, befouled, and degraded men out of malice prepense: but it was not so; like the miseducation of which we were talking just now, it came of their dreadful poverty. They were obliged to put up with everything, and even pretend that they liked it, whereas we can now deal with things reasonably, and refuse to be saddled with what we do not want.

Morris's last recorded reflections on Devon Great Consols came months before his death. Wilfred Scawen Blunt, his wife's lover, recorded in his diary on 29 May 1896:

> He told me something of his origin. His father was a bill broker in the City, and he himself was destined for that trade. If I had gone on with it, he said I should have broken the bills into very small bits. We had some mining shares in Cornwall, and when I succeeded to them I sold them. My relations thought me both wicked and mad, but the shares are worth nothing now.

It was not Morris's geography that was at fault – Devon Great Consols was classed as a Cornish mine, even though it was in Devon – it was his chronology. From succeeding to his shares in Devon Great Consols, to selling the last of them, took 22 years.

Morris failed to confess to his reading public that he himself had overseen and profited from the production of between one-third and a half of the world's arsenic. Arsenic that made its way into virtually every home in the land. Arsenic that polluted the Tamar. Arsenic that killed children in their beds, and girls in millinery factories. Arsenic used in medicines that gave patients skin cancer. Arsenic that poisoned the poor workers that Morris employed and the clientèle whose riches paid them. Arsenic that coloured his wallpapers green.

o-o

During Devon Great Consols' arsenic peak, medics turned their attention to the health impacts of production at the mine complex. To counteract arsenic toxicity men pushed a small wad of cotton wool into each nostril and wound handkerchiefs or scarves around their mouths. They smeared Fuller's earth on the exposed skin of their face, neck and hands. Photographs show workers handling huge amounts of white arsenic powder with no other protection. Miners and grinders were least at risk. Furnace operators suffered mainly when the wind direction caused them to inhale *smeech*, a smoke laden with arsenic and sulphur, resulting in extreme irritation of the respiratory tract.

Those that dug the arsenic out of the condensing baffles (Figure 7.3) suffered terribly. Many died from lung disease. More developed eruptions and pocks around the folds of the neck, nose, mouth, armpits and genitals, where arsenic dust dissolved in perspiration.

The worst job of all was removing vitriol solution from the depositing tanks of the copper works. These men occasionally breathed in arsine, a very potent toxin. They became dizzy and collapsed, and developed jaundice and bleeding kidneys.

Other symptoms of arsenic poisoning in miners included vomiting and abdominal cramps, conjunctivitis, swollen eyelids, pins and needles, and brown spots on the neck and temples. Most characteristic was a perforated nasal septum, as seen today in cocaine addicts.

The industry was brought to the attention of the Home Office and was then the subject of a joint inquiry by HM Deputy Inspector of Factories, Mr Gould, and HM Inspector of Mines, Mr Martin. Their *Report on Certain Alleged Cases of Poisoning in*

Figure 7.3 Removing arsenic from the chambers. From *Scientific American*, 1907. Reproduced with permission.

Arsenic Reduction Works was completed in 1901, but was never published. It was decided not to subject mining to rules similar to those governing the production of lead and arsenic paints. This was because arsenic was central to many areas of the British economy.

o–o

After Morris left the board and sold his remaining shares, Devon Great Consols limped on, but copper prices kept falling due to high-grade imports. The large stocks of arsenic had been disposed of as hoped. The year 1878 ran at a loss of £4,672 2s 6d. By 1879 the company was concentrating on arsenic, of which it was now the largest producer in the world. Large quantities of Cornish arsenic went to Germany for dye manufacture, and to the USA for controlling the Colorado beetle. Arsenic prices reached £20 per tonne, from 20 shillings a tonne a decade earlier. This also kept many tin mines going, and many new arsenic plants opened in south-west England. By 1880 Devon Great Consols declared its first dividend in three years. But the arsenic boom was over due to competition from Germany and Sweden.

Devon Great Consols as a company finally collapsed in 1902, having produced 73,000 tonnes of arsenic in 58 years, an estimated half of what was raised to the surface. The Duke of Bedford's estate quickly set about restoring the pheasant woods, selling off mine equipment, levelling the surface buildings, filling in the shafts and planting trees.

By 1904 there were only two major works left in the west of Cornwall. In 1906 there was a rapid revival in fortunes as US demand soared, but this was the final fling. In 1909 the US forbade the discharge of arsenic into the atmosphere and the massive US copper mines, such as that at Anaconda, had to filter out the element, causing US domestic production of arsenic to soar far beyond consumption.

In 1907 the price of arsenic had revived to £25 per tonne. The Duke instructed that arsenic production be reinstated at Blanchdown Wood. Mines were reopened and dumps reworked. The price dropped to £9 per ton by 1914, but the onset of war led to four more boom years. The British used arsenic to make the poison gas diphenylamine chloroarsine in answer to the less effective German mustard gas diphenyl chloroarsine. By 1918 arsenic cost £100 per tonne.

The interwar years saw an initial crash in prices, with subsequent fluctuations related to the need for pesticides to fight the cotton boll weevil until less dangerous pesticides were developed. A new chimney was erected at Blanchdown Wood in 1922 for the processing of arsenic, which can still be seen today.

With the Second World War came the development of chlorovinyldichloroarsine, or blister gas, and old mines were surveyed in preparation for mass production of the agent. But gas warfare in the Second World War turned out to be minimal and the arsenic reserves were not needed. Arsenic production at Blanchdown Wood ceased after the war.

o-o

While the English arsenic industry ground to an ignominious end during the 20th century, the USA manufactured a vast amount of the poisonous element, most of it as a by-product of the copper industry.

The Anaconda copper mines in Deer Lodge Valley, Montana, was the Devon Great Consols of the 20th century, the world's largest producer of arsenic. During 1883–84 the Anaconda Mining Company opened a copper reduction works. A new smelter, built in 1902, processed 7,000 tonnes of ore per day, but shut down in 1903 under a storm of protest from local farmers claiming that it was poisoning crops and livestock. Anaconda ended up paying $330,000 in compensation, as the 40 metre stacks of the smelter were emitting thousands of tonnes of arsenic. The company's solution was to build a stack 180 metres tall.

Eventually government regulations forced Anaconda and other companies to scrub arsenic trioxide from smelter gases, and the trioxide so produced became the world's main source of arsenic. The sale of arsenic from US smelters began in 1910. Production peaked in the USA in 1944 when 24,878 tonnes of arsenic were manufactured – a third of Devon Great Consols' entire output.

In 1985 arsenic production in the USA ceased owing to ever-tightening environmental laws. But the country's craving for arsenic has not abated – over 20,000 tonnes were imported in 1990.

o-o

What was the USA doing with all this arsenic? Spreading it liberally on the countryside. During the first half of the 20th century 90% of the element was turned into pesticides, particularly calcium arsenate, to fight the cotton boll weevil. Fruit crops were sprayed with lead arsenate and Paris green; sheep and cattle were dipped in sodium arsenite. In the 1970s the new organic arsenic compounds became popular as herbicides, particularly for cotton and grape cultivation: monosodium methylarsenate (MSMA), disodium methyl arsonate (DMSA) and dimethyl arsinic acid (cacodylic acid). MSMA and DMSA are the only remaining agricultural uses of arsenic. These compounds are also widely used on lawns and golf courses in doses of up to 100 kg of arsenic per acre, permanently contaminating domestic and recreational land. One study found up to 250,000 ppb of arsenic in the private gardens of Denver, Colorado, USA.

The extent of the problems caused by arsenic in US agriculture is outlined in Rachael Carson's epoch-making 1962 book *Silent Spring*:

> In arsenic-sprayed cotton country of southern United States, beekeeping as an industry has nearly died out. Farmers using arsenical dusts over long periods have been afflicted with chronic arsenic poisoning; livestock have been poisoned by crop sprays and weed killers containing arsenic. Drifting arsenic dusts from blueberry lands have spread over neighbouring farms, contaminating streams, fatally poisoning bees and cows, causing human illness.

Carson quotes Dr W.C. Hueper, an authority on environmental toxicology at the National Cancer Institute:

> It is scarcely possible... to handle arsenicals with more utter disregard of the general health than that which has been practised in our country in recent years. Anyone who has watched the dusters and sprayers of arsenical insecticides at work must have been impressed by the almost supreme carelessness with which the poisonous substances are dispensed.

The environmental and human health consequences of so much arsenic floating around the environment were and are dire.

Respiratory, bladder, liver and skin cancers caused by arsenic were rife in copper smelting, pesticide manufacture, gold mines and on farms. One study showed that 29% of men employed in manufacturing arsenical sheep dips died of cancer, compared to 13% of those in other occupations living in the same area.

A remarkable illustration of the lax US attitude towards arsenic is illustrated by rice cultivation. In the cotton states, such as Arkansas, the 1960s saw a drive to replace cotton production with rice. Rice planted in former cotton fields produced very little grain because of the amount of arsenic in the soil. In response farmers bred rice to produce high yields on contaminated soil.

In 2003, unable to find data on the arsenic content of rice grown on arsenic-rich soil, I decided to conduct some experiments for myself. I took American arsenic-sensitive and arsenic-resistant rice varieties and grew them on arsenic-contaminated soil. The arsenic levels in both sensitive and resistant rice grain of arsenic-treated plants doubled or trebled compared to plants grown on uncontaminated soil. Studies with pigs suggest that the body may readily assimilate arsenic in rice. Putting profit before human health did not end with in the 19th century, it seems.

o-o

Arsenic even played a role in the Vietnam War. The spraying of an area housing up to 4 million civilians with nearly 50 million litres of Agent Orange, a mixture of the dioxin-laced organic defoliants 2,4-D and 2,4,5-T is rightly infamous. But other pesticides were also used in this war: Agents Pink, Green, Purple, and White and 4.7 million litres of Agent Blue, all named after the coloured labels on their storage drums. Agent Blue is a blend of the organic arsenic compounds cacodylic acid and sodium cacodylate. No large-scale studies have been carried out to gauge the health consequences of this spraying of Vietnamese villages.

As agricultural use declined in the USA and elsewhere, another use of arsenic came to the fore. Wood treatment accounted for 80% of US demand in the 1990s. Wood impregnated with a toxic solution called copper chrome arsenic or CCA is less prone to rot when used outdoors in fence posts, decking and garden furniture. The treatment slows down decay, but does not stop it; so eventually the arsenic ends up in the soil. The composting or burning of CCA-treated wood also causes environmental contamination.

By the end of 1999, a grand total of 3.3 million tonnes of arsenic had been produced worldwide. Production proceeds. In 2000 alone, 50,000 tonnes was manufactured globally, mainly as a by-product of copper mining. Only part of the arsenic raised to the Earth's surface ends up as arsenic trioxide. Most is simply dumped around mines and ore-processing plants. South-east Asian (tin and copper), North and South American (gold and copper) and Australian (gold) mines have caused extensive arsenic pollution. At Getchel, Nevada, the horizon is filled with mountains of rock that has been ground, roasted and mixed with cyanide to extract gold. The newly exposed rock is streaked red with realgar and yellow with orpiment. Presumably the desert wastes of Nevada are deemed so vast that despoiling them on a grand scale is of little consequence.

o-o

Hueper in his book *Occupational Tumours* (1942) lists the industries in which arsenic-related skin lesions have been observed:

> ...glass workers; chemical workers (dye manufacturers, manufacturers of sheep dip); tanners; taxidermists; workers in lead factories; workers in rotogravure establishments (by the use of arsenic containing ink); workers in oil refineries (using an arsenic containing clay); workers in oil-cloth factories; in plants producing coloured papers; photographers (using

> arsenicals); seamstresses and weavers (handling goods dyed with arsenicals); workers in pelt and hair factories; agricultural workers; sprayers and dusters of arsenical insecticides; workers peeling fruit contaminated by arsenicals in canneries; artificial flower makers; arsenic roasters; carroters of felt hats; chargers in zinc smelters; color makers; colored-paper workers; compounders of rubber; copper founders; copper smelters; curriers in tanneries; cut-glass workers; decorators of pottery; dye makers; electroplaters; enamellers; ferro-silicon workers; fur handlers and preparers; galvanizers; gardeners; glass mixers; glaze dippers and mixer in pottery plants; gold refiners; insecticide makers; japan makers; jewellers; lead smelters; linoleum color workers; lithographers; mixers of rubber mordanters; paper glazers; paper hangers; pencil makers (working with colours); pitch workers; pottery workers; press-room workers in rubber; printers; pyrite burners; refiners of metals; rubber tire builders and workers; sealing-wax makers; sheep-dip makers; shot makers; sprayers of trees; sulphur burners; sulphuric acid workers; taxidermists; tinners; toy-makers; velvet makers; wall-paper printers; wax ornament workers; wire drawers, wood preservers; zinc mixers.

The natural calamity of poisoned tubewell water is one thing, but that we continue to spread arsenic around the environment deliberately, knowing all that the last two centuries have taught, is a gross form of folly.

Many arsenic products can be replaced with other compounds or elements; indeed, there has been a trend since the 1900s to do so. Environmental arsenic exposure from industry is decreasing, and hopefully will continue until arsenic is used only in products for which there are no alternatives (such as drugs for cancer and other diseases), or in products where it is biologically safe, like glass and semi-conductors.

As for mining and smelting, the only way to decrease pollution from these activities is to conduct them in a more environmentally aware manner. Modern metal processing has a lot more pollution control imposed on it, at least in developed nations, than in the past. But decreasing the impact of mining throughout the globe is costly. The real question is: are we willing to pay more for precious and base metal products?

Chapter 8

THE EXTRAORDINARILY PROTRACTED PROCESS

Can I please have some more arsenic in my water, Mommy?

US Democratic National Committee commercial

Enormous tracts of North and South America have arsenic-rich groundwater. The largest continuous contaminated area in the world is the Pampean Plain in Argentina. At over 1,000,000 square kilometres it is six times the size of Bangladesh. The arsenic-contaminated Antofagasta region of Northern Chile is the size of just one Bangladesh. Bangladesh would also fit snugly into the arsenic-rich aquifer of Arizona, USA.

Arsenic-induced skin disease in Chile and Argentina was first officially recorded in 1917, in the journal *Círculo Médico del Rosario*. The lesions were dubbed 'Bell Ville' after the town in Córdoba, Argentina, where skin pigmentation and cancers were common. In the 1970s a study published in the *Archives of Argentinean Dermatology* found high levels of lung and bladder cancers in the same place.

Bangladeshis, in the main, have only been exposed to arsenic for the last 30 years. Indigenous South Americans have been drinking it for around 10,000 years. They suffer much the same diseases, including black rain and warts.

There was some medical evidence that native South Americans have adapted to cope. Humans, like most mammals, metabolise arsenic into the organic compounds monomethyl arsonic acid (MMA) and dimethylarsinic acid (DMA). No one knows exactly why. It may be because MMA and DMA are less toxic and can be readily excreted as urine. The biochemical mechanisms behind this conversion are also not well understood. Arsenic in most human urine is typically 10–30% inorganic arsenic, 10–20% MMA and 60–80% DMA. In 1990s a Swedish toxicologist Marie Vahter found that Andean women native to high arsenic areas metabolise and excrete the poison differently from recent immigrants. The women's urine has just 2% MMA and around 75% DMA, Vahter found. She hypothesised that this shift has a genetic basis.

In the mid-1990s the belief that people indigenous to arsenic-rich regions are somehow resistant was fuelled by preliminary research published by Allan Smith and colleagues at the University of California. It hinted that some populations in Chile's Atacama region did not have a high incidence of arsenic-related disease. Speculation about genetic adaptation was published in the journal *Science* in 1998 in an article entitled 'Toxicologists shed new light on old poisons'.

But, also in 1998, Smith revisited the subject with a detailed study of 400,000 people in the city of Antofagasta, published in the *American Journal of Epidemiology*. He estimated that 7% of the population over the age of 30 would die from arsenic-induced disease, lung and bladder cancers. It was clear that genetic adaptation to arsenic had not occurred in this region.

Water supplies in this area have been cleaned since the 1970s, so the incidence should drop in younger generations. Nonetheless, Smith's article ends with a chilling statement:

> High priority should be given to studies at lower levels of exposure such as those associated with 50 ppb, the drinking water standard for much of the world.

Including Bangladesh.

Smith also found that well-fed inhabitants of the village of Chiu Chiu in the Atacama were highly prone to arsenic-induced disease, dispelling another arsenic myth: that its effects are exacerbated by poor diet. This is often believed to be a major factor in Bangladesh.

Unlike the Bengal Basin, it is river waters rather than aquifers that feed arsenic to the residents of northern Chile. Records, started around the 1950s, show that arsenic levels were high in a number of towns, ranging from 60 to 600 ppb. These concentrations either stayed constant, or, as in the case of Antofagasta, rose by up to 10-fold until the 1970s. In the mid-70s water treatment plants were installed in large conurbations. The village of San Pedro was too small to benefit from such technology; in 1994 it still had arsenic levels of 600 ppb.

o-o

The geology, topology and climate of America's arsenic-affected aquifers could not be more different from those of the wet, low-lying Bengal Basin. The arsenical areas of South America include some of the driest places on earth. No rain has fallen in recorded history on parts of Chile's Atacama Desert. The desert borders the 4,000 metre Altoplano, which itself borders the Andes. The area is busy with volcanic activity, thought to provide the arsenic. Indeed, it gushes out of the earth at hot springs and geysers. Elsewhere on the Altoplano and the Pampean Plain older volcanic rock is lower in arsenic. Here the desiccated climate concentrates the poison in groundwaters. Unlike the neutral, anaerobic aquifers of Bangladesh and West Bengal, America's aquifers tend to be alkaline, saline and well stocked with oxygen.

The arid conditions of Chile and Argentina are mirrored in Mexico, New Mexico, Arizona, California and Nevada. Here too, aquifers are salty and highly alkaline. Here too, water tables can have arsenic at up to 2,600 ppb. Of the 20,044 potable groundwater supplies tested by the United States Geological Survey between 1976 and 2000, 2% had arsenic levels greater than 50 ppb (see Figure 8.1). The current official US safe level is 10 ppb. Twelve per cent of US drinking water supplies exceed this. The average in Nevada is 31 ppb. Averages over 10 ppb are found in North Carolina, North Dakota, New Hampshire, Montana, Delaware and Arkansas. In 2001 the US Environmental Protection Agency estimated that 12.7 million Americans – that's 4.3% of the population – currently drink tap water with over 10 ppb of arsenic each day.

o-o

Back to Allan Smith. In 2002 Smith and his co-workers published a critique in *Science* of what they called the 'extraordinarily protracted process' of arsenic standard setting in the USA. Basically it

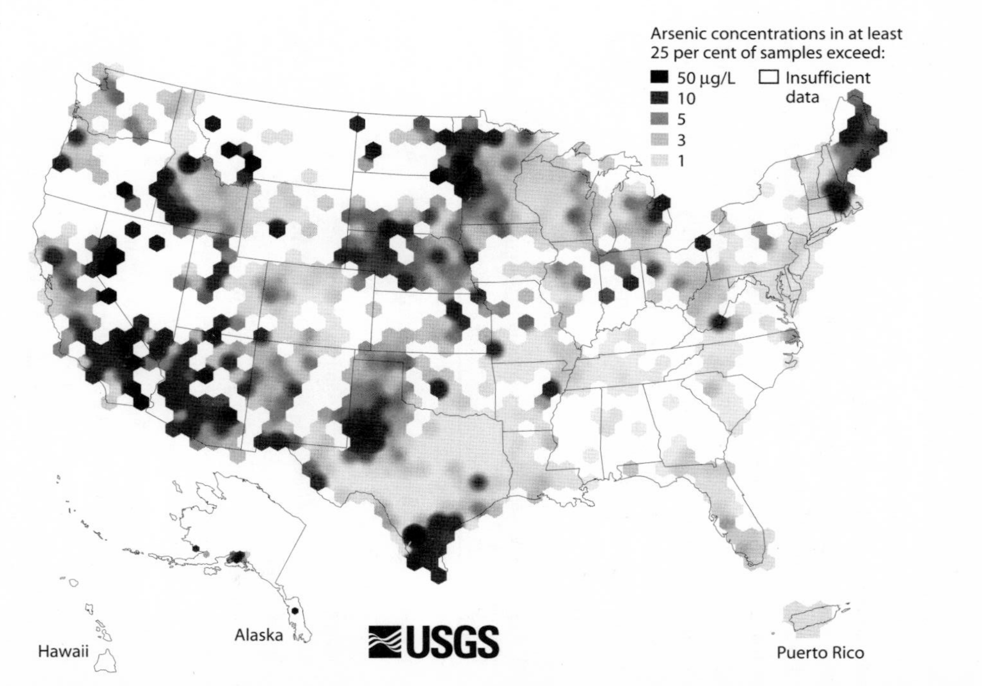

Figure 8.1 United States Geological Survey map of arsenic levels in aquifers utilised for drinking water. Note that the unit µg/L is equivalent to ppb.

has been a cat and mouse game, the latest instalment of which, notoriously, was one of the very first acts of George W. Bush's presidency. He stalled the lowering of 'safe' arsenic levels authorised by the Clinton administration just two months previously. This vacillation, as illustrated in Table 8.1, has scotched any coordinated response to arsenic crises in the USA and beyond.

Admittedly, quantifying the cancer risk posed by arsenic is hard. The lab animals normally used to test carcinogens are poor analogues for human cancers where arsenic is concerned. So the risk for humans has been estimated from studies of populations exposed to arsenic in groundwaters, primarily in Taiwan and Chile. As with any human epidemiology study, these feature variables and uncertainties open to statistical interpretation. How

Table 8.1 History of US standards for arsenic in drinking water, adapted from *Science*, 296, pp. 2145–6.

1942	US sets drinking water standard of 50 ppb
1962	US identifies 10 ppb as its goal
1975	EPA adopts standard of 50 ppb
1986	Congress directs EPA to revise standard by 1989
1988	EPA estimates that the ingestion of 50 ppb in drinking water causes a cancer risk of 1 in 400
1992	Internal cancer risk estimated to be 1.3 per 100 persons at 50 ppb by EPA
1993	WHO recommends lowering limit to 10 ppb
1996	Congress directs EPA to propose a new standard by 2000
2000	EPA proposes a standard of 5 ppb, requesting comment on 3, 10 and 20 ppb
2001	(January) Clinton EPA moves to lower the standard to 10 ppb
2001	(March) Bush EPA delays lowering the standard
2001	(September) New NRC report concludes that EPA underestimated cancer risks
2001	(October) EPA announces it will adopt the standard of 10 ppb
2001	WHO does not set an arsenic standard in EHC document
2002	(February) Effective date for the new standard of 10 ppb
2006	Compliance date for new standard

long people have been drinking contaminated water differs from study to study, for example, as do arsenic levels, age, illness, nutrition, smoking, genetics, gender and so on. Some even argue that the Taiwanese and Chilean studies cannot be extrapolated to the USA at all, as diets are so different.

But as Smith put it in his *Science* article:

> ...when there is direct human epidemiological evidence that a substance causes cancer, we should focus on margins of safety, avoiding extensive statistical manipulations of data and excessive debate about potential uncertainties. Prudent public health decisions should not wait until there is proof of serious risks at low exposure.

In short, the dangers are too great to waste time. Low limits need to be set to ensure that we are not poisoned by what comes out of our taps.

o–o

So why did George W. Bush get so controversially interested in arsenic? Some have suggested that his presidential campaign was heavily supported by mining lobbies – mining being a major source of arsenic pollution in the USA. Whether or not this is the case, a 10 ppb limit has enormous financial implications for the US water industry. The US Environmental Protection Agency (EPA) estimates that the national annual cost of a 10 ppb arsenic standard will be approximately $181 million. However, the American Water Works Association Research Foundation, in a report issued in October 2000, estimated that the cost of compliance for the new standard would be $590 million annually with an initial capital cost of $4.5 billion.

Bush defended his move at a press conference on 29 March 2001, two months after he was sworn in:

> QUESTION: Mr. President, in the last few weeks you have rolled back health and safety and environmental measures proposed by the last administration, and other previous administrations. This has been widely interpreted as a payback time to your corporate donors. Are they more important than the American people's health and safety? And what else do you plan to repeal?
>
> THE PRESIDENT: I told people pretty plainly that I was going to review all the last-minute decisions that my predecessor had made, and that is exactly what we're doing. I presume you're referring to the decision on arsenic in water. First of all, there had been no change in the arsenic – accepted arsenic level in water since the '40s. And at the very last minute, my predecessor made a decision, and we pulled back his decision so that we can make a decision based upon sound science and what's realistic.
>
> There will be a reduction in the acceptable amount of arsenic per billion after the review in the EPA.

Bush's EPA administrator, Christine Todd Whitman, said that the Clinton administration's standard had not been based on the 'best available science'. Over three years the Clinton administration spent $2.5 million refining arsenic standard setting. Clinton tasked the US National Research Council (NRC) to sift through the available data and advise accordingly. The NRC's first findings were published in a document called *Arsenic in Drinking Water* in 1999. It concluded:

> Upon assessing the available evidence, it is the subcommittee's consensus that the current EPA MCL [maximum concentration limit] for arsenic in drinking water of 50 ppb does not achieve EPA's goal for public health protection and therefore requires downward revision as promptly as possible.

Table 8.2 Theoretical Maximum-Likelihood Estimates, as incidence per 10,000 people, for arsenic-derived cancer above the baseline cancer rate, over a lifetime of arsenic exposure.

Arsenic concentration (ppb)	Bladder cancer		Lung cancer	
	Females	Males	Females	Males
3	4	7	5	4
5	6	11	9	7
10	12	23	18	14
20	24	45	36	27

A 2001 revision of this report clearly set out the cancer risks of arsenic in drinking water (see Table 8.2).

Former Bush speech writer David Frum states in his book *The Right Man* that Bush's chief political advisor, Karl Rove 'pressed for reversal' of the arsenic standard to win votes in New Mexico, where a 10 ppb standard would mean higher taxes.

The delay in arsenic standard setting caused many to ridicule Bush's already weak environmental credentials. 'This decision will force millions of Americans to continue to drink arsenic-laced water', said Erik D. Olson of the National Resources Defence Council. 'Many will die from arsenic-related cancers and other diseases, but George Bush apparently doesn't care. This outrageous act is just another example of how the polluters have taken over the government'. Eventually, the Bush administration was forced to adopt the 10 ppb level that it had initially rejected. The scientific evidence was too strong.

While Bush has been lampooned for his environmental policies, Republicans can point out that for seven years, 11 months and 28 days in office, Clinton kept the arsenic levels where they had been since 1942, at 50 parts per billion. Only in the waning hours of his administration did he reduce the allowable concentration to 10 ppb.

o–o

The US government's attempt to balance the health of its citizens and their appetite for paying tax has had severe ramifications for the arsenic crises in the rest of the world.

WHO's International Programme on Chemical Safety (IPCS) produces risk assessments for chemicals, primarily as guidelines for developing countries. These assessments are called Environmental Health Criteria (EHC). The first EHC on arsenic was published in 1981. This concluded that there was too little evidence to set standards. By 2001, the world of arsenic had moved on. The risks were better understood as high levels were found in region after region.

The revised EHC, in which I was involved, was published in 2001, nearly two years after the meeting to finalise its text. All eyes were on the document: any discrepancy between a WHO standard and a US Environmental Protection Agency standard would have been embarrassing. Two members of the US EPA helped write the arsenic assessment. Three EPA personnel were on the secretariat overseeing the writing of the document. The US EPA also sponsored the National Research Council arsenic documents being prepared at the same time.

The EPA was keen to avoid the kind of confusion that surrounded dioxins. Dioxins are organic chlorine-containing pollutants, from organochlorine pesticides and chlorine containing chemical waste, which affect our immune system and hormones. In 1998 WHO set a 'tolerable daily intake' (TDI) of 10 picograms/kilogram/per day, which is a massive 1,670 times greater than the current US EPA standard of 0.006 picograms/kilogram/per day. Most countries around the world with regulations plumped for the WHO standard.

The revised EHC on arsenic is odd. It is devoid of recommendations for safe levels of arsenic in food and water – surely its *raison d'être*? The only allusion to standard setting is in the short section of the book entitled *Previous evaluations by international bodies*: 'The World Health Organization has given a provisional guideline value of 10 ppb for arsenic in drinking-water as the practical quantification limit (WHO, 1996)'. In other words, according to WHO, 10 ppb is not necessarily safe, it's just easily measurable. This in itself is misleading. I, along with other chemists, was routinely measuring 1 ppb, using relatively basic equipment, in 1989.

The statement about previous standard setting is strange too. A 1993 WHO document called *Guidelines for drinking-water quality, 2nd Edition* states: 'With a view to reducing the concentration of this carcinogenic contaminant in drinking-water, a provisional guideline value for arsenic in drinking-water of 10 ppb is established. The estimated excess lifetime skin cancer risk associated with exposure to this concentration is 6×10^{-4}'. To put it more clearly, six more people in every 10,000 will get skin cancer than would do so where levels are virtually zero. These cancer predictions are linear models, so the risk for drinking 50 ppb is $6 \times 5 = 30$ more cases of skin cancer in every 10,000 people. So, in 1993 WHO calculated that 3 extra people per 1,000 would get skin cancer if they regularly drank water with 50 ppb arsenic. Where was this in 1996 or 2001?

What the 2001 EHC does say is:

> Arsenic exposure via drinking-water is causally related to cancer in the lungs, kidney, bladder and skin. Drinking-water arsenic concentrations of ≤50 ppb have been associated with increased risk of cancer in the bladder and lung. Precursors of skin cancer have been associated with drinking-water arsenic levels of ≤50 ppb.

A statement of fact. What fails to follow is any WHO suggestion of what should be done about this fact.

So the most comprehensive risk assessments suggest that 50 ppb arsenic in drinking water is unsafe, yet this is the figure considered 'safe' by WHO's UNICEF in Bangladesh. Despite all this there is still no concrete plan to provide Bangladeshis with water at 50 ppb, never mind 10. The British Geological Survey estimates that 25% of tubewells have more than 50 ppb arsenic. If the safe level was 10 ppb, 45% of Bangladesh's tubewells would be unusable. That's nearly five and a half million wells. The practical implications of standard setting are enormous.

o-o

The 2001 EHC on arsenic has another strange omission: food. It is usual when calculating the risk posed by a cancer-causing chemical that a Maximum Tolerable Daily Intake (MTDI) is formulated, to account for all sources. WHO did this in its provisional 1993 assessment: it came up with a figure of 2 micrograms per kilogram of body mass per day. If a Bangladeshi weighing 60 kg drinks 3 litres of water containing 50 ppb arsenic, she is ingesting 2.5 micrograms per kilogram body mass per day, before eating. Rice, on average, naturally contains around 200 ppb arsenic. A 60 kg Bangladeshi eats around half a kilogram of rice per day. So rice will contribute 1.7 micrograms per kilogram body mass per day. Add this rice figure to the water figure and your 60 kg Bangladeshi is consuming over double WHO's maximum tolerable arsenic intake – more if her rice has taken up arsenic from tainted irrigation. In affected parts of Bangladesh I have recorded up to 1,800 ppb in rice samples, so a 60 kg person could ingest 8.3 micrograms of arsenic per kilogram per day – over four times the 1993 WHO maximum tolerable daily limit – before drinking a drop of water. Maize, the staple food of South Americans, also has high levels of arsenic, reaching up to 1,850 ppb, in the Andean villages of northern Chile.

The failure of WHO to take a clear stance has had a curious corollary. Regional regulatory bodies have taken their cue from Australia, the only country to put an upper limit on the amount of arsenic permissible in food: 1,000 ppb. Reports on arsenic levels in foods of Bangladesh claim they are safe because they fall below this 1,000 ppb level. Unfortunately, as the calculations above illustrate, 1,000 ppb is way off the mark. It was set because

Australians eat a lot of seafood that is naturally high in arsenic. But the arsenic in seafood is in forms with very low toxicity, mainly the compound arsenobetaine or 'fish arsenic'.

The type of arsenic in a foodstuff is of utmost importance in assessing the danger it poses. The arsenic in American rice is predominantly in the form of dimethylarsinic acid (DMA). DMA is a lot less toxic than inorganic arsenic species such as arsenate and arsenite – after all, it is the form we produce and excrete. Bangladeshi rice, regrettably, is quite different. Most of the arsenic is in the most toxic form of the element, arsenite. By a cruel variation in biology, the rice varieties grown in Bangladesh are the most dangerous to its population. Poisoned water, poisoned rice.

Chapter 9
JOI BANGLA!

They say a terrible hard thing who assert that the division of the world's production to afford each one a mouthful of food, a bit of clothing, is only a Utopian dream. All these social problems are hard indeed! Fate has allowed humanity such a pitifully meagre coverlet, that in pulling it over one part of the world, another has to be left bare. In allaying our poverty we lose our wealth, and with this wealth what a world of grace and beauty and power is lost to us.

Rabindranath Tagore, *Glimpses of Bengal*, Shelidah, 10 May 1893.

One in every sixty people on the planet is living in an area where they may be exposed to drinking water with 50 ppb arsenic or over. One hundred million people. Two or three times more may quaff water with 10 ppb arsenic or above.

This mass poisoning is covered occasionally by the more environmentally aware newspapers or television channels, but mostly there is silence. Unlike a famine or civil war, where the scale of human suffering is immediately obvious, arsenic is an insidious

Figure 9.1 *Death's dispensary* by George Pinwell, 1966, Philadelphia Museum of Art, illustrating the risk of microbial disease from London water. Tubewells in Bangladesh exchanged such deadly pathogens for arsenic. Reproduced with permission.

killer. Its victims die slowly in remote villages, hidden away from the media glare.

For the nations who have the resources to deal with their tainted groundwater, action has been or is being taken. Legislation, at long last, is in place so that all US citizens should have safe water by 2006. Poorer Chile has spent the past 30 years working to purify contaminated supplies.

In the parts of the world only recently discovered to be affected by arsenic, such as Vietnam, Nepal, Pakistan, Iraq, China, Laos and Myanmar, research is still required to assess the scale of the problem. In West Bengal, where the crisis has been known for some 20 years, remedial action has been slow and disappointing.

For the worst hit country, Bangladesh, the problem is accelerating. The most effective action taken here so far has been the testing of tubewells for arsenic. But new wells are sunk faster than old ones can be checked, especially now that private enterprise is in on the act.

When in 1998 the World Bank gave the government of Bangladesh a $34 million interest-free loan to solve the problems, this led to the founding of the Bangladesh Arsenic Mitigation and Water Supply Project, based in Dhaka and run by the Government with aid agencies and World Bank assistance. But uncertainty over what mitigation technologies would best benefit Bangladesh menat that just over one sixth of the loan was spent by 2004. Of the further $55 million now pledged, $40 million is from the World Bank, and the remainder from government, private sector and community contributions.

A sum of $89 million over 8 years sounds like a lot, but it works out at just over a dollar for each of the 80 million or so people at risk from arsenic in Bangladesh. The net income into the World Bank for the financial year 2002 was $2,778 million. And compared to money spent on other environmental disasters? Exxon has paid out $5 billion to clean up the mess caused by the 1989 *Exxon Valdez* oil spill in Prince William Sound, Alaska. The accident killed, it is estimated, 300 harbour seals, 2,800 sea otters,

250 bald eagles and maybe 22 killer whales. In 2004 it was discovered that private wells supplying 70 homes in Miami, Florida had arsenic levels between 52 and 81ppb. Total cost of cleaning up the well water: $750,000 – over $10,000 dollars per household. The City of Scottsdale, Arizona, population 203,000, has a $64 million budget over 18 months to meet the new US 10 ppb standard.

An analysis of the 2004 Net Official Development Assistance or Official Aid Figures published by the World Bank shows that Bangladesh is high on the list of aid receivers, getting $1.2 billion yearly. But it is far from being the highest. China, Egypt, India, Israel, Indonesia, Poland and the Russian Federation all receive similar or greater aid. The per capita spend on aid in Bangladesh is just $8 per year. Albanians get $86, Bosnians $160, Hondurans $103, Macedonians $122, Nicaraguans $178 and Serbians $123. Of course aid is needed in other parts of the globe besides Bangladesh, and all the countries mentioned have suffered greatly. Fortunately, the average income is low in Bangladesh, as are living expenses, so every extra dollar goes further than in most parts of the globe. Nonetheless, much more aid is now required to fight arsenic.

o-o

The infrastructure needed to tackle the Bangladesh catastrophe is simply not there; it would take decades to build, even if the will and resources were available. Bangladesh has less than 3,000 miles of railway, one telephone for every 500 people and one car for each 2,500. According to the World Bank in 2004 'nearly 70% of the population does not have access to electricity, while those with access face power cuts on a regular basis'. Each hospital in Bangladesh services between 1 and 2 million people.

For a few of the very worst affected villages UNICEF and other aid agencies have installed clean water supplies. But as The World Bank put it 2004: 'The unprecedented nature, enormity, and complexity of Bangladesh's problem... along with a lack of sufficient coordination of efforts so far, have caused serious delays in arsenic mitigation progress from the national to the local level'.

The scientists involved in surveying wells and searching for alternative supplies of safe water are highly committed. Their research is vital to the future of Bangladesh and other nations. Lots of technologies have been explored; all have limitations. There is no miracle solution. In UNICEF's appraisal:

> There is no technological magic bullet that will solve the arsenic problem quickly and easily. Proper application of technology in the context of Bangladesh's financial, institutional and infrastructural constraints is more difficult – and ultimately more important – than the technology itself. UNICEF will continue to focus on identifying and supporting low-cost appropriate alternatives and arsenic removal technologies that are chosen, accepted and managed by communities.

o–o

In Bangladesh over 10 million tubewells need to be tested for arsenic. From 1998 to date, at least 1.3 million water samples have been analysed using field kits. One million pumps have been painted red for poisoned or green for safe, in an effort funded by the World Bank and UNICEF.

The most reliable way to test for arsenic is to take a water sample back to the lab and analyse it by atomic absorption spectrometry (AAS) or atomic fluorescence spectroscopy (AFS) hooked up to a hydride-generator. The hydride-generator converts soluble arsenic to the gas arsine (AsH_3). The amount of arsine produced is measured by how much light is emitted (AFS) or absorbed (AAS) when wavelengths of specific energy are shone through a cell in which the arsine is heated until it decomposes. AAS or AFS machines cost $10,000 to $20,000 dollars, are robust and cheap to run and can detect arsenic levels as low as

100 parts per trillion. These machines can tackle about 20 samples per hour, or less than 200 samples per working day. To test 10 million wells with one machine would take 10,000 weeks – 192 years. Alternatively, 192 machines could run 10 million samples in a year. In other words, it's very costly. On top of that there are the logistics of getting samples safely from the field to the laboratory and communicating the results back. These calculations do not even take into account that wells need re-testing, and that new wells are being sunk all the time.

Because of these difficulties of large-scale lab testing, government departments, NGOs and aid agencies favour field test kits. These are simple chemical assays where reagents mixed with water react to cause a colour change. In theory, the intensity of the colour is related to the concentration of arsenic in the water. A task force trained to use these kits is methodically marching over Bangladesh.

Most of the tests are based on the Gutzeit method. A strong solution of stannous chloride, potassium iodide, powdered zinc and hydrochloric acid converts any inorganic arsenic to the arsine gas. The arsine then reacts with a strip of paper impregnated with mercuric bromide, turning it from white to yellow or brown, depending on how much gas is made. The colour is matched with a reference swatch by eye.

These field tests sound good in principle, but in practice are fraught with pitfalls. The first is their sensitivity. A standard test kit manufactured by the chemical company Merck was generally used before 2000. This strip could measure accurately only down to 100 ppb. Since 2000 more sensitive kits that detect levels of 10 ppb arsenic have been available from Merck and others. The second problem is climate. Intense heat and humidity can render the chemicals ineffective over time. Then there is user variation. The assays rely on the skill of the tester to gauge any colour change. Normal differences in testers' visual acuity and competence give rise to variation in their results. Finally, arsenic levels in tubewells are not constant; there is a natural fluctuation of about 15–20% over a year. Thus wells that have concentrations around the so-called safe level of 50 ppb may one year be painted green but the next year daubed red, confusing villagers no end.

Dipankar Chakraborti's research team surveyed the efficacy of field test kits by comparing their results with those of a hydride-generation/AAS set up. The field test kits gave false negatives (saying that the tubewell water was safe when it was not) in up to 68% of cases, and false positives (saying that clean or low-risk water was unsafe) up to 35% of the time. The main source of error was small colour changes at the limits of the method's sensitivity. Chakraborti conducted this study under laboratory conditions, using trained scientists. The uncertainties can only increase in the field. Chakraborti estimates that 45% of tubewells tested by field kits are inappropriately labelled. Of those painted green, 7.5% are actually unsafe, he believes.

Leaving aside for a moment the drawbacks of testing, let's turn to the way in which wells are taken out of action. They are painted. If villagers have to travel distances of hundreds of metres or more to obtain 'green' water they can be tempted to use the red pumps. Destroying wells would be the best way to inactivate them.

One short-term solution is tubewell switching. Within contaminated regions well water is often highly heterogeneous. So those living near a red well could simply be encouraged to use a green one a little further a field. A detailed study of the Araihazar administrative district in Bangladesh measured arsenic levels in nearly 5,000 tubewells. It found that 90% of inhabitants were within 100 metres of a well with arsenic levels below 50 ppb. There may be some social barriers to switching, given the strong caste structure in the region. 'If I die, I will die, but I will not go to fetch water from another man's house', one prosperous farmer said, despite his disfiguring arsenicosis. Also, women, who collect most of the water, are not supposed to leave their bari (a cluster of related households) unaccompanied. If tubewells are near latrines, privacy is also an issue.

o–o

Identifying contaminated tubewells is one thing; giving the citizens of the Bengal Delta clean water is quite another.

The first option is to return to dug wells, the traditional water supply before tubewells. In 1960, when dug wells predominated, more than 1 in 7 infants and almost 1 in 4 children under 5 died of water-borne diseases such as cholera and dysentery. Dug wells tap only the surface of an aquifer. In theory water can be treated with ultraviolet light or chlorinated to kill human pathogens. But both technologies require considerable infrastructure – not least a reliable supply of electricity, chemicals and regular expert maintenance. There are around 80,000 villages in Bangladesh. Maintaining water treatment plants for so many sites is nigh-on impossible.

Others have suggested that shallow dug wells could be covered or surrounded with concrete or iron to minimise contact with disease carriers such as human and animal effluent. But it only takes one breach for the well to be contaminated.

Another approach is to sterilise well water with potassium permanganate. This too has downsides. Potassium permanganate leads sludge to build up in a well, blocking the pump. And it does little to control coliform bacteria that cause gastrointestinal illness, particularly after monsoons.

Ultimately, the risk of microbial disease from dug wells has to be balanced against the risk of arsenic-induced illness from shallow tubewells. Currently, water-borne pathogens kill up to 1 in 20 infants and 1 in 10 under fives in Bangladesh each year. WHO predicts that in one of the worst affected villages in Bangladesh, Samta, up to 6.5% of adults could eventually develop cancer due to arsenic. As a 2003 WHO report *Arsenic, Drinking-water and Health Risks Substitution in Arsenic Mitigation* concluded: 'the risk posed by microbial hazards is greater than for arsenic.... This does not imply that arsenic mitigation is not important, but to emphasise the need for emergency response measures to ensure that risks from microbial hazards do not increase'. The people of the Bengal Basin have a choice: the frying pan or the fire.

One technology to remove hazardous microbes, widely used in rural communities around the world, is sand or gravel filtration. Microorganisms move much more slowly through tightly packed sand and gravel than in free water. Thus if filter beds are regularly maintained, unsafe dug well water can be transformed, cheaply and readily, into clean water.

Such systems have to be resilient enough to cope with yearly floods. Plus, if broken or poorly maintained, they quickly contaminate water themselves, releasing their rich store of pathogens. No wonder studies on recently constructed dug wells and sand filters found villagers very reluctant to use them. In 2004 Karin Kemper, World Bank Senior Water Resources Management Specialist in Bangladesh, explained:

> Small-scale implementation of arsenic mitigation or prevention options, such as promoting the use of commonly-shared wells which have tested safe for arsenic, actually represent a step backwards in the context of Bangladesh. The rural population has become used to a high service level in terms of privately-owned hand pumps in their yards.

UNICEF is just as wary. Its report of 1999 contains the following on returning to shallow dug well technology:

> Hastily and indiscriminately switching back to surface water across the country without strong pollution control measures in place will most likely do more harm than good to the people of Bangladesh. A balanced package of interventions, that uses safe surface, ground and rain water sources as dictated by the local situation will ultimately be the most appropriate solution.

o-o

If Bengal's yearly deluge of rainwater could be caught and stored correctly on a large scale, it might offer a way out. The water is simple enough to harvest. Corrugated or plastic roofs have runoff collection barrels. When the rains start, the first few barrels are discarded, as these are chock full of dust, insects and bird droppings. The next portion of water is cleaner. Storage is the tricky part. Harvested water has to be protected from insects and animals over the long hot months of the dry season. To do this in any quantity would require tanks, a mechanism to keep the water pure and regular inspections.

The simplest option might seem to be household purification, either by using some sort of filtration device or simply sprinkling chemicals into water. Between the late 1990s and 2003 this was widely touted as the short-term solution to Bangladesh and West Bengal's woes. Aid and World Bank money flowed into developing the technology and providing propaganda to help its uptake. The 'three pitchers' purification system is the most widely advertised. Here water is poured from one earthenware pitcher into another. The first contains iron chips and coarse sand. The second features charcoal and fine sand; the third collects the filtered water. Arsenic binds to the iron chips and charcoal, purifying the water. But the villagers didn't like it, for lots of reasons. It is time-consuming and generates toxic waste. Plus, getting a continuous supply of iron chips and charcoal to 80 million people, when road and rail connections are patchy at best, is no mean feat.

There is one potential long-term solution. The vast majority of tubewells more than 200 metres deep have low arsenic levels. This must be to do with the geology of the different layers of the aquifer. While the Holocene strata are rich in arsenic, the deeper Pleistocene sediments appear not to be.

Deeper tubewells have not caught on. They cost 45 times more to build than their shallow cousins, making them, in the main, a communal resource, with all the implications that holds for women carrying water over considerable distances. One survey of households that shifted to safe water supplies found that the average distance travelled to a well increased from under 30 metres to 170 metres.

More importantly, Bengalis are now suspicious of tubewells, shallow or deep. I visited Samta, where seven deep tubewells have been sunk to provide an alternative water supply to the poisoned shallow tubewells. Villagers rarely drink from these new wells, even though they are safe; the residents prefer their central sand filter unit. In a village so scarred by the calamity, tubewells equate to danger.

And rightly so. Even geologists and hydrologists do not want to give the go-ahead for a massive program to sink deep tubewells. They fear that these too could eventually become tainted, creating even more problems down the line. It is essential that we understand the fundamental mechanisms that make groundwater high in arsenic before assuming that currently safe water will remain so.

o-o

The World Bank now considers that the best plan for Bangladesh is to pipe water from where it is plentiful and clean to where it is not. In 2003 the Bank summed up its position thus:

> One of the most promising approaches, now being tried on a pilot basis, is the establishment of rural piped water schemes. These can match the convenience of the now contaminated household tubewells, and allow for better control of water quality testing and treatment at the community level. Such systems, however, require increased levels of community organization and technical input and demand the active collaboration of national and local governments, civil society, and the private sector.

Water supply specialist Kemper again:

> In embarking on this innovative approach to provide piped water, the Government is taking a bold step towards creating new institutional arrangements in the sector and, at the same time, promoting decentralization of the provision of services to the local levels. We hope this project will help the government increase coverage and quality of water supply, in particular to the poorer populations, which is essential to increased health and productivity in the country.

The World Bank and others are sinking $89 million into piped water and central treatment facilities.

o–o

Public perception is central to fighting this crisis. Water is such a fundamental part of our existence that any major cultural changes to its use need to be explained very carefully. Tubewell water tastes and looks fine and doesn't give people diarrhoea. Arsenic poisoning from the same water is imperceptibly slow. Too slow to make people jump into action.

Local government departments have conducted poster campaigns, featuring diagrams for a largely illiterate population, to educate citizens in the most affected areas. These generally warn people away from dangerous wells with red symbols.

Are they working? A survey, *Fighting arsenic, listening to rural communities*, conducted by WHO in 2001 visited over 2,400 households. Only 87% of respondents in arsenic-affected areas knew of the problems; awareness was at 53% in unaffected areas. Most respondents did not know of the serious health effects caused by the arsenic in their drinking water – 42% continued to use contaminated tubewells because there were no alternative sources. The main option was to switch to public tubewells, which invariably would mean a longer walk to get water. A tiny proportion – 2.5% of those surveyed – had changed to pond or tank water, enduring the additional expense and time required to boil the water.

Asked about preferred alternatives to tubewell water, including six arsenic mitigation technologies, 76% were prepared to pay for arsenic-free water. Roughly 72% preferred community-based schemes, while 28% wanted household purification. More than 70% wanted deep tubewell water, while dug wells came lowest. But in the worst-afflicted places all tubewells are spurned in favour of filter units. Here terrible suffering has given the residents a rather different perspective.

o-o

For the ill and dying of Bangladesh, little help is available. Arsenicosis victims are given antioxidants, such as vitamin C, and a clean water supply if they are lucky. There is only rudimentary medical infrastructure, even in the hardest-hit areas.

Take Samta again, the place that journalists, NGOs, and aid agencies are taken to see. The village itself is quite a surprise. It looks like a rural idyll. Beautifully kept mud homes nestle among trees; tidy yards and lush vegetable patches are tended by villagers in colourful saris – not what the centre of the 'world's worst chemical disaster' is supposed to look like. Wander around, though, and the horror reveals itself. Villagers show you keratosed feet, bodies splattered with black rain, four-fingered hands and scars where malignant skin cancers have been removed. Around 250 people in this village of 5,000 have arsenic-induced disease, from young mothers with babies in their arms to old men and women. Twenty or so have already died.

What do they make of the constant intrusion into their world? They are infuriated that it has left them with scant tangible care.

o-o

Humanity's past and present are bathed in arsenic. It polluted our workplaces, poisoned our living rooms, and caused cancers from cure-alls. Arsenic contaminates vast tracts of the globe through the exploitation of naturally rich aquifers, the dumping of mining waste, the application of pesticides and the preservation of timber. Food is grown on arsenic-contaminated soil and tens of millions drink arsenic-contaminated water. Mankind has ignored the dangers.

Things are dire in Bangladesh at the moment, but there is hope. International aid agencies and NGOs now at least agree that arsenic in groundwaters constitutes a disaster of enormous proportions. UNICEF and the World Bank are committed to piping safe water into homes or communal standpipes – an excellent solution. If they pull this off it will be the organisational wonder of the 21st century. Let's hope they do. The total cost of the piped water plan has not been calculated; it will likely be thousands of millions of dollars. Successful solutions pioneered in Bangladesh will have to be replicated elsewhere – to bring clean water to the arsenic-threatened inhabitants of south-east Asia.

Meanwhile, no aid agency should support the indiscriminate sinking of tubewells into alluvial sediments in south-east Asia. Such wells will have a high probability of being arsenical. Only aquifers that have been thoroughly investigated and proven safe should be considered for the supply of domestic and agricultural water.

Will arsenic be the death of Bangladesh? Absolutely not. This is a country born into strife and raised on resilience. Dhaka seems to be germinating the seeds of an economic miracle, like those witnessed in other mega-cities of south-east Asia. Of course, Bangladesh will trade on its very cheap labour. But just maybe the resulting wealth will for once filter back down, and with support from the outside world, overcome 'the devil's water'. Let's hope the Bangladesh liberation cry '*Joi Bangla!*' – 'Long Live Bengal!' rings true.

FURTHER READING

Chapter 1: The devil's water

O'Rourke, P.J. (1994) *All the Trouble in the World: The Lighter Side of Famine, Pestilence, Destruction and Death*. Picador, London.

Pilger, J. (2001) *Heroes*. Vintage, London.

Shand, M. (2003) *River Dog: Travels down the Brahmaputra*. Abacus, London.

Chapter 2: A natural disaster

Frankenberger Jr, W.T. (ed.) (2002) *Environmental Chemistry of Arsenic*. Marcel Dekker, New York.

Welch, A.H. and Stollenwerk, K.G. (eds.) (2003) *Arsenic in Ground Water: Geochemistry and Occurrence*. Kluwer Academic Publishers, Boston.

Chapter 3: Fools' gold

Gregory, A. (2001) *Eureka!: The Invention of Science*. Icon Books, London.

Jacobi, J. (1958) *Paracelsus: Selected Writings*. Pantheon Books, New York.

Strathern, P. (2000) *Mendeleyev's Dream: The Quest for the Elements*. Hamish Hamilton, London.

Turner, H.R. (1995) *Science in Medieval Islam*. University of Texas Press, Austin.

Chapter 4: The verdant assassin

Ball, P. (2001) *Bright Earth: The Invention of Colour*. Viking, London.

Bristow, I.C. (1996) *Interior House-Painting Colours and Technology 1615–1840*. Yale University Press, New Haven.

MacCarthy, F. (1994) *William Morris*. Faber, London.

Chapter 5: Healing arsenic

Marquardt, M. (1949) *Paul Ehrlich*. William Heinemann, London.

Chapter 6: To frustrate the aim of justice

Altick, R.D. (1970) *Victorian Studies in Scarlet: Murders and Manners in the Age of Victoria*. Norton & Company, New York.
Bell, G. (2002) *The Poison Principle*. Macmillan, London.
Blyth, H. (1975) *Madeleine Smith*. Duckworth, Unwin Brothers Limited, Surrey.
Hartman, M.S. (1976) *Victorian Murderesses*. Schocken Books, New York.
Watson, K. (2004) *Poisoned Lives: English Poisoners and their Victims*. Hambledon & London, London.

Chapter 7: Nothing green met the eye

Barton, D.B. (1970) *Essays in Cornish Mining History*, Vol. 2. Barton Limited, Penzance.
Carson, R. (1965) *Silent Spring*. Penguin, London.
Harvey, C. and Press, J. (1996) *Art, Enterprise and Ethics: The Life and Works of William Morris*. Frank Cass, London.
Hueper, W.C. (1942) *Occupational Tumors and their Allied Diseases*. Charles C. Thomas, Springfield.

Chapter 8: The extraordinarily protracted process

National Research Council (1999) *Arsenic in Drinking Water*. National Research Council, Washington DC.
National Research Council (2001) *Arsenic in Drinking Water: 2001 Update*. National Research Council, Washington DC.
WHO (2001) *Arsenic and Arsenic Compounds*, 2nd edn. Environmental Health Criteria 224. WHO, IPCS, Geneva.

INDEX